Advances in Water Distribution Networks

Advances in Water Distribution Networks

Special Issue Editors

Giuseppe Pezzinga
Enrico Creaco

MDPI • Basel • Beijing • Wuhan • Barcelona • Belgrade

MDPI

Special Issue Editors
Giuseppe Pezzinga
Dipartimento di Ingegneria Civile e Architettura
Università di Catania
Italy

Enrico Creaco
Dipartimento di Ingegneria Civile e Architettura
Università di Pavia
Italy

Editorial Office
MDPI
St. Alban-Anlage 66
4052 Basel, Switzerland

This is a reprint of articles from the Special Issue published online in the open access journal *Water* (ISSN 2073-4441) from 2017 to 2018 (available at: https://www.mdpi.com/journal/water/special_issues/distribution_network).

For citation purposes, cite each article independently as indicated on the article page online and as indicated below:

LastName, A.A.; LastName, B.B.; LastName, C.C. Article Title. *Journal Name* **Year**, *Article Number*, Page Range.

ISBN 978-3-03897-556-4 (Pbk)
ISBN 978-3-03897-557-1 (PDF)

Contents

About the Special Issue Editors

Giuseppe Pezzinga obtained a PhD in Hydrodynamics in October, 1990. His career developed at the University of Catania as researcher from 1989 to 1992, as an associate professor from 1992 to 2000, and as a full Professor of Hydraulics since 2000. He has taught fluid mechanics (from 1992 to 2010) and computational hydraulics (from 1993 to 2011). He currently teaches advanced hydraulics. He has been the tutor of more than 110 degree theses. He has taught PhD courses at the University of Catania. He was the Director of the second Master Course in "Water Management and Environment Protection", organized by the Istituto Superiore di Catania per la Formazione di Eccellenza (2002–03). He was Chairman of the Didactic Council of Civil Engineering of the University of Catania from 2006 to 2012. He conducts research in the field of theoretical and applied hydraulics, working mainly on secondary currents in pressure and free surface flows; unsteady friction in pipe flows; protecting devices in unsteady flows; flood propagation models; models for blood flow in the arterial system; and the optimization of hydraulic networks. He is author of more than 70 scientific papers presented in international and national journals and conferences. The contributions were mainly improvements of theoretical models and numerical schemes, together with experimental research carried out for validation. He has been the manager of research financed by the University of Catania, and the local manager of research financed by the Italian Ministry of the University. He has been a referee for several international journals.

Enrico Creaco obtained his PhD in Hydraulic Engineering in 2006 and has researched topics pertinent to water and environmental systems for over ten years. His career began at the Universities of Catania and then Ferrara, Italy. From May 2014 to June 2015 he took up a Research Fellow post at the University of Exeter and became Assistant Professor at the University of Pavia in September 2015. He has been Associate Professor at the University of Pavia since September 2018. Since 2016, he has been Honorary Senior Research Fellow at the University of Exeter and Adjunct Senior Lecturer at the University of Adelaide. In September 2018, he obtained the Full Professor Qualification in the Hydraulics, Hydrology and Hydraulic Infrastructure Sector. He has lectured on hydraulic infrastructures at both undergraduate and postgraduate level and have published more than 50 papers in a variety of Scopus- and ISI-indexed international journals. He is Associate Editor of the Journal of Water Resources Planning and Management-ASCE and has participated in/coordinated various national and international research projects. His research interests are the design and management of water distribution, sewer and irrigation systems; numerical modelling of shallow water and sediment transport; demand analysis; and the protection of water distribution systems from contamination.

Preface to "Advances in Water Distribution Networks"

The research into water distribution networks has recently been characterized by a deep renewal and development, due to technical progress in control systems and computational resources. In recent decades, the research has examined in depth the well-established topics related to the quantitative simulation and optimization of water distribution systems, and it has broadened to include water quality aspects, such as those concerning the network capacity in terms of residual disinfection and its protection from the effects of accidental or terroristic contamination events.

Giuseppe Pezzinga, Enrico Creaco
Special Issue Editors

water

MDPI

Editorial

Advances in Water Distribution Networks

Enrico Creaco [1] and Giuseppe Pezzinga [2,*]

[1] Dipartimento di Ingegneria Civile e Architettura, University of Pavia, Via Ferrata 3,
     27100 Pavia, Italy; creaco@unipv.it
[2] Dipartimento di Ingegneria Civile e Architettura, University of Catania, Via Santa Sofia 64,
     95123 Catania, Italy
*   Correspondence: giuseppe.pezzinga@unict.it; Tel.: +39-095-738-2708

Received: 4 October 2018; Accepted: 25 October 2018; Published: 30 October 2018

Abstract: This Editorial presents a representative collection of 10 papers, presented in the *Special Issue on Advances in Water Distribution Networks* (WDNs), and frames them in the current research trends. Four topics are mainly explored: simulation and optimization modelling, topology and partitioning, water quality, and service effectiveness. As for the first topic, the following aspects are dealt with: pressure-driven formulations, algorithms for the optimal location of control valves to minimize leakage, benefits of water discharge prediction for the remote real time control (RTC) of valves, and transients generated by pumps operating as turbines (PATs). In the context of the second topic, a topological taxonomy of WDNs is presented, and partitioning methods for the creation of district metered areas (DMAs) are compared. With regards to the third topic, the vulnerability to trihalomethane is assessed, and a statistical optimization model is presented to minimise heavy metal releases. Finally, the fourth topic focusses on estimation of non-revenue water (NRW), inclusive of leakage and unauthorized consumption, and on assessment of service under intermittent supply conditions.

Keywords: water distribution networks; non-revenue water; leakage; energy; real time control; pumps as turbine; pressure-driven analysis; topology; partitioning; district metered areas; water quality; trihalomethane; heavy metals

1. Introduction

The research on water distribution networks (WDNs) has recently undergone important renewal and development, due to technical progress in control systems and computational resources. In the last few decades, the research has examined in depth the well-established topics related to the quantitative simulation and optimization of water distribution systems, and it has broadened to water quality aspects.

As for water quantity, one of the most explored topics is undisputedly modelling [1]. In this context, the extended period simulation, which represents WDN behaviour as a sequence of steady states, is by far the most widely adopted modelling tool. This is because it enables obtaining good results in the trade-off between accuracy and computation burden. However, the thorough assessment of nodal outflows is an essential requirement for the accuracy of the simulation. To this end, the pressure-driven approach originally proposed by Bhave [2], which relates nodal outflows to demands and pressure-heads, proved to perform much better at reproducing WDN behaviour in a wide range of service pressure, in comparison with the demand-driven approach, in which nodal outflows are set equal to demands. However, little attention has been dedicated to the comparison of the various pressure-driven formulations available in the scientific literature, e.g., [2–7].

Besides outflow to authorized users, nodal outflows also include non-revenue water (NRW) [8], which is represented by leakage and unauthorized consumption. Numerous studies, e.g., [9–11], have been carried out on the effective parameters for characterizing WDN in terms of NRW. However, the correlation between NRW ratio to overall water production and WDN operational and physical parameters has not been thoroughly explored.

To decrease the NRW ratio, as well as to obtain other benefits such as the reduction in pipe bursts and the extension of infrastructure life, water utility managers often choose to perform service pressure regulation in WDNs. This requires the installation of control valves at suitable locations and the real time control (RTC) of these devices, to meet the demand variations in time. After the work of Jowitt and Xu [12], many other works, e.g., [13–17], in the scientific literature explored optimal location of control valves in WDNs. Nevertheless, few comparative works exist to help water utility managers choose the best algorithm for generic case studies. As for RTC, some work was done to set-up increasingly effective algorithms to regulate control valve settings as a function of the pressure at critical nodes [18] and variables, such as water discharges at valve sites [19]. An interesting idea, which can be further developed, was recently presented by Page et al. [20,21] to include water discharge forecast within the control logic.

Instead of being dissipated at control valves, the surplus of service pressure can be recovered by installing turbines or pumps operating as turbines (PATs) [22,23]. While the performance of these devices has been analysed in many contexts, e.g., [24–27], including real WDNs, e.g., [28,29], little attention has been given so far to the effects of transients generated by turbines or PAT.

Besides service pressure regulation, WDN management includes other practices such as partitioning. This is done to subdivide the WDN into sufficiently small areas, called district metered areas (DMAs) [1], to facilitate management and monitoring. Numerous algorithms, e.g., [30–35], are proposed in the scientific literature for WDN partitioning, while few comparative analyses are available. Furthermore, since WDN partitioning is generally carried out by making use of the graph theory, a study is missing on the behaviour of WDNs from the topological point of view, to characterize their basic metrics.

While the research on WDNs in the 20th century was mainly dedicated to water quantity aspects, the two first decades of the current century have seen the birth of new research lines concerning water quality. These lines have received nourishment from the changes in the regulations in many countries, which encourage the draft and adoption of water safety plans [36]. A water safety plan is a plan to ensure the safety of drinking water using a comprehensive risk assessment and risk management approach, which encompasses all steps in water supply from catchment to user. In this context, the WDN plays an important role since it can be threatened by events of accidental and intentional contaminations [37–39]. These events can cause supplied water to contain inacceptable concentrations of undesired compounds, such as heavy metals, disinfection by products and so forth. Therefore, methodologies for improving water quality and estimating the vulnerability of WDNs to these compounds should be developed.

2. Overview of the *Special Issue*

The *Special Issue* was established to point out the recent trends on WDNs, with emphasis on the opportunities introduced by technical progress for simulation, design, and management of water distribution systems. The collected papers are representative of some current main research topics in WDNs.

2.1. WDN Simulation and Optimization

Four papers form part of this topic:

- Ciaponi and Creaco [40] present the comparison of five pressure-driven formulations in the context of WDN modelling. The results of two case studies show that the formulations tend to behave similarly in terms of nodal outflows. The formulations with smooth relationship between nodal outflow and pressure head tend to guarantee faster algorithm convergence, in comparison with a relationship with derivative discontinuities. The results yielded by the formulations for low values of the nodal pressure head can be very different;

- Creaco and Pezzinga [41] present the comparison of two different algorithms for the optimal location of control valves for leakage reduction. The former is based on the sequential addition (SA) of control valves: at each step, the optimal combination of valves is searched for, while containing the optimal combination found at the previous step. Therefore, the former algorithm searches for only one new valve location at each step, among all the remaining ones. The latter algorithm consists of a multi-objective genetic algorithm (GA), in which valve locations are encoded inside individual genes. The results obtained on two WDNs show that SA and GA yield identical results for small number of valves. When this number grows, GA performs increasingly better than SA. However, the smaller computation time of SA may make this algorithm preferable in the case of large WDNs;
- Creaco [42] explores the benefits of water discharge prediction in the RTC of WDNs. An algorithm aimed at controlling the settings of control valves and variable speed pumps, as a function of pressure head signals from remote nodes in the network, is used. Two variants of the algorithm are considered, based on the measured water discharge in the device at the current time and on the prediction of this variable at the new time, respectively. The RTC algorithm attempts to correct the expected deviation of the controlled pressure head from the set point, rather than the currently measured deviation. The results of the applications prove that RTC benefits from the implementation of the prediction, in terms of closeness of the controlled variable to the set point;
- Pérez-Sánchez et al. [43] characterize the water hammer phenomenon in the design of PAT systems, emphasizing the transient events that can occur during a normal operation. This is based on project concerns towards a stable and efficient operation associated with the normal dynamic behaviour of flow control valve closure or by the induced overspeed effect. The analysis shows how precise evaluation of basic operating rules depends upon the system and component type, as well as upon the required safety level during each operation, with emphasis on the analysis of transients.

2.2. WDN Topology and Partitioning

Two papers belong to this topic:

- Giudicianni et al. [44] apply Complex Network Theory to characterize the behaviour of WDNs from a topological point of view. A tool of analysis is provided to help in finding solutions to several problems of WDNs. The application of the methodology to 21 existing networks and 13 literature networks highlights some topological peculiarities and the possibility to define a set of best design parameters for ex-novo WDNs. Also, the interplay between topology and some performance requirements of WDNs is discussed;
- Liu et al. [45] present a comparative analysis of three partitioning methods, including Fast Greedy, Random Walk, and Metis, which are commonly used to establish the DMAs in water distribution systems. A complex water distribution network is used for comparison considering two cases, i.e., unweighted and weighted edges, where the weights are represented by the demands. The results obtained from the case study network show that the Fast Greedy method is more effective in the weighted graph partitioning. The study provides an insight for the application of the topology-based partitioning methods to establish district metered areas in a water distribution network.

2.3. Water Quality

Two papers are concerned with water quality issues:

- Quintiliani et al. [46] propose a methodology for estimating the vulnerability with respect to users' exposure to disinfection by-products (DPBs) in WDNs. The presented application considers total trihalomethane (TTHM) concentrations, but the methodology can be used also for other types of DPBs. Five vulnerability indexes are adopted. The results obtained on five case studies suggest that the introduced indexes identify different critical areas in terms of elevated concentrations of TTHMs. This allows identification of the higher risk nodes in terms of different kinds of exposure (short period of exposure to high TTHMs values, or chronic exposure to low concentrations);

- Peng and Mayorga [47] propose a statistical multiple objectives optimization, namely Multiple Source Waters Blending Optimization (MSWBO), to find optimal blending ratios of source waters for minimizing three heavy metals (HMR) in a WDN. Three response surface equations are applied to describe the reaction kinetics of HMR, and three dual response surface equations are used to track the standard deviations of the three response surface equations. A weighted sum method is performed for the multi-objective optimization problem to minimize three HMRs simultaneously. The experimental data of a pilot distribution system are used to demonstrate the model's applicability, computational efficiency, and robustness.

2.4. Service Effectiveness

Two papers consider service effectiveness aspects:

- Jang and Choi [48] estimate the NRW ratio, that is the ratio of losses from unbilled authorized consumption and apparent and real losses to the total water supply. NRW is an important parameter for prioritizing the improvement of a WDN. The paper shows that the accuracy of multiple regression analysis (MRA) is low compared to the measured NRW ratio, where the accuracy of estimation by an artificial neural network (ANN) with the optimal number of neurons, is higher;
- Mokssit et al. [49] propose a methodology for assessing the effectiveness of water distribution service in the context of intermittent supply, based on a comparison of joint results from literature reviews and feedback from drinking water operators who had managed these networks, with standards for defining the effectiveness of drinking water service. The results are used to structure an evaluation framework for water service and to develop improvement paths defined in intermittent networks. The resulting framework highlights the means available to water stakeholders to assess their operational and management performance in achieving the improvement objectives defined by the environmental and socio-economic contexts in which the network operates. Practical examples of intermittent system management are collected from water system operators and presented for illustration purposes.

3. Discussion

All the papers of the special issue are focussed on topics that are at the forefront of the research in WDNs. Besides achieving the expected objectives, they are founded on very accurate reviews of the most recent works in the scientific literature.

Four papers in the special issue namely [40,41,45,49] have the merit of presenting comparative analyses. The results of these works may become a benchmark for future reference and may also offer precious information to orientate future research efforts.

Another merit of the papers in the special issue lies in the fact that the methodology proposed are applied to real WDNs, considering complete or skeletonized layouts. This confers reliability to the results obtained.

The methodologies adopted are multi-faceted, ranging from hydraulic [40–43] and water quality [46,47] modelling, to the graph theory [44,45], statistical [42,48], and optimization [41,47] techniques.

However, the special issue feels the effect of a strong and widespread tendency in the scientific literature, which has recently been producing many more numerical models and mathematical methods than experimental studies for model validation. In fact, only one of the ten papers, namely [43], reports novel experimental results. The others, instead, use literature data for model validation. It is the Authors' idea that the collection of new data from laboratory and in-situ experiments, by means of modern and accurate equipment available nowadays, could enable more accurate validation of numerical models and mathematical methods developed so far.

4. Conclusions

The paper presents an overview on the present research topics on WDNs through the analysis of the literature and of the papers presented in the *Special Issue on Advances in Water Distribution Networks*. The analysis of existing literature is carried out to put in evidence the aspects covered by the research in WDNs. With regards to water quantity, one of the most explored topics is the modelling, both for simulation and for optimisation purposes. Attention is focused on: demand models, pressure driven formulation, hypotheses on flow. In the context of WDN simulation, a paper presented in the special issue compares various pressure driven formulations [40]. With regards to optimisation, although there is a wide consideration of aspects to be optimised, in recent years the regulation of pressure by valves and the recovery of energy surplus by micro-turbines or PATs have assumed a fundamental role. As for WDN optimization, a paper in the special issue presents the comparison of deterministic and probabilistic algorithms for the location of control valves [41]. Another paper of the special issue shows the extent to which RTC of valves can benefit from implementation of water discharge prediction at valve site [42]. Staying in the context of WDN modelling, a paper presented in the special issue concerns energy recovery by means of PATs, with emphasis on the analysis of transients [43].

Along with service pressure regulation, another commonly adopted practice in WDN management is partitioning, which consists of WDN subdivision into small areas, which can be easily monitored and managed. A topic very closely related to partitioning is topology. Numerous algorithms were proposed in the scientific literature for WDN partitioning, generally carried out by making use of the graph theory. A paper presented in the special issue compares WDN partitioning algorithms [45]. Another paper proposes a topological taxonomy of WDNs [44].

The research lines considering water quality aspects have received recent attention from the adoption of water safety plans, to ensure safety of drinking water. Methodologies are currently being developed to improve water quality and to estimate the vulnerability of WDNs, to avoid the risk of contamination of supplied water by undesired substances. As for water quality, two papers were presented in the special issue, concerning assessment of vulnerability to trihalomethane [46] and development of a statistical optimization model to improve drinking water quality through the minimization of heavy metal releases [47], respectively.

Aspects related to service effectiveness have also received increased attention. This is because water authorities need to save water resources and to optimize their financial resources, while meeting users' satisfaction. Besides traditional issues, such as those associated with guaranteeing the suitable service pressure at users' connections, subjects as NRW, and intermittent supply conditions have recently been explored. In the context of the analysis of service effectiveness, the special issue includes two papers. The former concerns estimation of NRW, inclusive of leakage and unauthorized consumption, using multiple regression analysis and artificial neural networks [48]. The latter, instead, proposes a methodology for assessing service effectiveness under intermittent supply conditions [49].

Each of these papers gives contributions on the research in WDNs and gives possible research topics to be developed in the future.

Funding: This research received no external funding.

Acknowledgments: This research was conducted using the funds supplied by the University of Pavia and by the University of Catania.

Conflicts of Interest: The author declares no conflict of interest.

References

1. Walski, T.M.; Chase, D.V.; Savic, D.A.; Grayman, W.; Beckwith, S.; Koelle, E. *Advanced Water Distribution Modeling and Management*; Haestad Methods, Inc.: Waterbury, CT, USA, 2004.

2. Bhave, P.R. Node flow analysis of water distribution systems. *J. Transp. Eng.* **1981**, *107*, 457–467.

3. Wagner, B.J.M.; Shamir, U.; Marks, D.H. Water distribution reliability: Simulation method. *J. Water Resour. Plan. Manag.* **1988**, *114*, 276–294. [CrossRef]

4. Fujiwara, O.; Li, J. Reliability analysis of water distribution networks in consideration of equity, redistribution, and pressuredependent demand. *J. Water Resour. Res.* **1998**, *34*, 1843–1850. [CrossRef]

5. Tucciarelli, T.; Criminisi, A.; Termini, D. Leak Analysis in Pipeline System by Means of Optimal Value Regulation. *J. Hydraul. Eng.* **1999**, *125*, 277–285. [CrossRef]

6. Tanyimboh, T.; Templeman, A. Seamless pressure-deficient water distribution system model. *J. Water Manag.* **2010**, *163*, 389–396. [CrossRef]

7. Ciaponi, C.; Franchioli, L.; Murari, E.; Papiri, S. Procedure for defining a pressure-outflow relationship regarding indoor demands in pressure-driven analysis of water distribution networks. *Water Resour. Manag.* **2015**, *29*, 817–832. [CrossRef]

8. Frauendorfer, R.; Liemberger, R. *The Issues and Challenges of Reducing Non-Revenue Water*; Asian Development Bank: Mandaluyong, Philippines, 2010.

9. Jung, J.J. The Primary Factor of Management Evaluation Indicators for Local Public Water Supplies & Suggestion of Alternative Evaluation Indicators. *J. Korean Policy Stud.* **2012**, *12*, 139–159.

10. Lambert, A.O.; Brown, T.G.; Takizawa, M.; Weimer, D. A Review of Performance Indicators for Real Losses from Water Supply Systems. *J. Water SRT Aqua* **1999**, *48*, 227–237. [CrossRef]

11. Shinde, V.R.; Hirayama, N.; Mugita, A.; Itoh, S. Revising the Existing Performance Indicator System for Small Water Supply Utilities in Japan. *Urban Water J.* **2013**, *10*, 377–393. [CrossRef]

12. Jowitt, P.W.; Xu, C. Optimal Valve Control in Water-Distribution Networks. *J. Water Resour. Plan. Manag.* **1990**, *116*, 455–472. [CrossRef]

13. Reis, L.; Porto, R.; Chaudhry, F. Optimal location of control valves in pipe networks by genetic algorithm. *J. Water Resour. Plan. Manag.* **1997**, *123*, 317–326. [CrossRef]

14. Araujo, L.; Ramos, H.; Coelho, S. Pressure control for leakage minimisation in water distribution systems management. *Water Resour. Manag.* **2006**, *20*, 133–149. [CrossRef]

15. Pezzinga, G.; Gueli, R. Discussion of "Optimal Location of Control Valves in Pipe Networks by Genetic Algorithm". *J. Water Resour. Plan. Manag.* **1999**, *125*, 65–67. [CrossRef]

16. Creaco, E.; Pezzinga, G. Embedding Linear Programming in Multi Objective Genetic Algorithms for Reducing the Size of the Search Space with Application to Leakage Minimization in Water Distribution Networks. *Environ. Model. Softw.* **2015**, *69*, 308–318. [CrossRef]

17. Covelli, C.; Cozzolino, L.; Cimorrelli, L.; Della Morte, R.; Pianese, D. Optimal Location and Setting of PRVs in WDS for Leakage Minimization. *Water Resour. Manag.* **2016**, *30*, 1803–1817. [CrossRef]

18. Campisano, A.; Creaco, E.; Modica, C. RTC of valves for leakage reduction in water supply networks. *J. Water Resour. Plan. Manag.* **2010**, *136*, 138–141. [CrossRef]

19. Creaco, E.; Franchini, M. A new algorithm for the real time pressure control in water distribution networks. *Water Sci. Technol. Water Supply* **2013**, *13*, 875–882. [CrossRef]

20. Page, P.R.; Abu-Mahfouz, A.M.; Yoyo, S. Parameter-Less Remote Real-Time Control for the Adjustment of Pressure in Water Distribution Systems. *J. Water Resour. Plan. Manag.* **2017**, *143*, 04017050. [CrossRef]

21. Page, P.R.; Abu-Mahfouz, A.M.; Mothetha, M.L. Pressure Management of Water Distribution Systems via the Remote Real-Time Control of Variable Speed Pumps. *J. Water Resour. Plan. Manag.* **2017**, *143*, 04017045. [CrossRef]

22. Kougias, I.; Patsialis, T.; Zafirakou, A.; Theodossiou, N. Exploring the potential of energy recovery using micro hydropower systems in water supply systems. *Water Util. J.* **2014**, *7*, 25–33.

23. Pérez-Sánchez, M.; Sánchez-Romero, F.; Ramos, H.; López-Jiménez, P.A. Energy Recovery in Existing Water Networks: Towards Greater Sustainability. *Water* **2017**, *9*, 97. [CrossRef]

24. Ramos, H.M.; Borga, A.; Simão, M. New design solutions for low-power energy production in water pipe systems. *Water Sci. Eng.* **2009**, *2*, 69–84.

25. Carravetta, A.; Del Giudice, G.; Fecarotta, O.; Ramos, H.M. Energy Recovery in Water Systems by PATs: A Comparisons among the Different Installation Schemes. *Procedia Eng.* **2014**, *70*, 275–284. [CrossRef]

26. Caxaria, G.; de Mesquita e Sousa, D.; Ramos, H.M. Small Scale Hydropower: Generator Analysis and Optimization for Water Supply Systems. 2011, p. 1386. Available online: http://www.ep.liu.se/ecp_article/index.en.aspx?issue=57;vol=6;article=2 (accessed on 12 March 2017).

27. Sinagra, M.; Sammartano, V.; Morreale, G.; Tucciarelli, T. A new device for pressure control and energy recovery in water distribution networks. *Water* **2017**, *9*, 309. [CrossRef]

28. Muhammetoglu, A.; Nursen, C.; Karadirek, E.; Muhammetoglu, H. Evaluation of performance and environmental benefits of a full-scale pump as turbine system in Antalya water distribution network. *Water Sci. Technol. Water Supply* **2017**, *18*, 130–141. [CrossRef]

29. Balacco, G.; Binetti, M.; Caporaletti, V.; Gioia, A.; Leandro, L.; Iacobellis, V.; Sanvito, C.; Piccinni, A.F. Innovative mini-hydro device for the recharge of electric vehicles in urban areas. *Int. J. Energy Environ. Eng.* **2018**. [CrossRef]

30. Clauset, A.; Newman, M.E.J.; Moore, C. Finding community structure in very large networks. *Phys. Rev. E* **2004**, *70*, 06611. [CrossRef] [PubMed]

31. Di Nardo, A.; Di Natale, M.; Santonastaso, G.; Tzatchkov, V.; Alcocer-Yamanaka, V. Water network sectorization based on graph theory and energy performance indices. *J. Water Resour. Plan. Manag.* **2013**, *140*, 620–629. [CrossRef]

32. Giustolisi, O.; Ridolfi, L. A novel infrastructure modularity index for the segmentation of water distribution networks. *Water Resour. Res.* **2014**, *50*, 7648–7661. [CrossRef]

33. Scarpa, F.; Lobba, A.; Becciu, G. Elementary DMA design of looped water distribution networks with multiple sources. *J. Water Resour. Plan. Manag.* **2016**, *142*, 04016011. [CrossRef]

34. Sela Perelman, L.; Allen, M.; Preis, A.; Iqbal, M.; Whittle, A.J. Automated sub-zoning of water distribution systems. *Environ. Model. Softw.* **2015**, *65*, 1–14. [CrossRef]

35. Di Nardo, A.; Giudicianni, C.; Greco, R.; Herrera, M.; Santonastaso, G. Applications of graph spectral techniques to water distribution network management. *Water* **2018**, *10*, 45. [CrossRef]

36. WHO (World Health Organization). *Guidelines for Drinking-Water Quality—First Addendum to Third Edition, Volume I: Recommendations*; WHO: Geneva, Switzerland, 2006.

37. Nilsson, K.A.; Buchberger, S.G.; Clark, R.M. Simulating Exposures to Deliberate Intrusions into Water Distribution Systems. *J. Water Resour. Plan. Manag.* **2005**, *131*, 228–236. [CrossRef]

38. Khanal, N.; Buchberger, S.G.; McKenna, S.A. Distribution system contamination events: Exposure, influence, and sensitivity. *J. Water Resour. Plann. Manag.* **2006**, *132*, 283–292. [CrossRef]

39. Davis, M.J.; Janke, R. Importance of Exposure Model in Estimating Impacts When a Water Distribution System Is Contaminated. *J. Water Resour. Plan. Manag.* **2008**, *134*, 449–456. [CrossRef]

40. Ciaponi, C.; Creaco, E. Comparison of Pressure-Driven Formulations for WDN Simulation. *Water* **2018**, *10*, 523. [CrossRef]

41. Creaco, E.; Pezzinga, G. Comparison of Algorithms for the Optimal Location of Control Valves for Leakage Reduction in WDNs. *Water* **2018**, *10*, 466. [CrossRef]

42. Creaco, E. Exploring Numerically the Benefits of Water Discharge Prediction for the Remote RTC of WDNs. *Water* **2018**, *9*, 961. [CrossRef]

43. Pérez-Sánchez, M.; López-Jiménez, P.A.; Ramos, H.M. PATs Operating in Water Networks under Unsteady Flow Conditions: Control Valve Manoeuvre and Overspeed Effect. *Water* **2018**, *10*, 529. [CrossRef]

44. Giudicianni, C.; Di Nardo, A.; Di Natale, M.; Greco, R.; Santonastaso, G.F.; Scala, A. Topological Taxonomy of Water Distribution Networks. *Water* **2018**, *10*, 444. [CrossRef]

45. Liu, H.; Zhao, M.; Zhang, C.; Fu, G. Comparing Topological Partitioning Methods for District Metered Areas in the Water Distribution Network. *Water* **2018**, *10*, 368. [CrossRef]

46. Quintiliani, C.; Di Cristo, C.; Leopardi, A. Vulnerability Assessment to Trihalomethane Exposure in Water Distribution Networks. *Water* **2018**, *10*, 912. [CrossRef]

47. Peng, W.; Mayorga, R.V. Developing a Statistical Model to Improve Drinking Water Quality for Water Distribution System by Minimizing Heavy Metal Releases. *Water* **2018**, *10*, 939. [CrossRef]

48. Jang, D.; Choi, G. Estimation of Non-Revenue Water Ratio Using MRA and ANN in Water Distribution Networks. *Water* **2018**, *10*, 2. [CrossRef]

49. Mokssit, A.; de Gouvello, B.; Chazerain, A.; Figuères, F.; Tassin, B. Building a Methodology for Assessing Service Quality under Intermittent Domestic Water Supply. *Water* **2018**, *10*, 1164. [CrossRef]

water

MDPI

Article

Comparison of Pressure-Driven Formulations for WDN Simulation

Carlo Ciaponi and Enrico Creaco *

Dipartimento di Ingegneria Civile e Architettura, Univ. of Pavia, Via Ferrata 3, 27100 Pavia, Italy;
ciaponi@unipv.it (C.C.)
* Correspondence: creaco@unipv.it

Received: 24 March 2018; Accepted: 19 April 2018; Published: 21 April 2018

Abstract: This paper presents the comparison of five pressure-driven formulations in the context of water distribution network (WDN) modelling. These formulations, which relate nodal outflow q to users to demands d and nodal pressure heads h, were implemented inside the global gradient algorithm for the snapshot solution of the equations concerning mass and energy conservation at WDN nodes and pipes, respectively. The modelling of leakage nodal outflows as a function of pressure was also considered. The applications concerned two case studies, in which nodal demands were suitably amplified to lower service pressure below the desired values. This was done to stress the effects of the pressure-driven dependence $q(h)$ in the WDN. The results showed that the formulations tend to behave similarly in terms of nodal outflows. Compared to a widely used formulation, which features a $q(h)$ relationship with derivative discontinuities, the other four formulations analyzed tend to guarantee faster algorithm convergence, above all for simple and poorly interconnected WDNs, due to their smooth $q(h)$ relationship. The results in terms of nodal pressure heads can be very different, above all for low values of h.

Keywords: water distribution network; snapshot simulation; pressure-driven

1. Introduction

The traditional approach for simulating water distribution networks (WDNs) is the demand-driven approach (DDA) [1–4], in which nodal outflows are set equal to demands independently from service pressure. In this context, the solution of mass and energy balance equations enables determining nodal heads and pipe flows in either snapshot or extended-period simulations [5]. The DDA yields satisfying results in some applications, such as WDN design, in which service pressure is assumed to be larger than the desired value for full demand satisfaction. However, some abnormal operational scenarios [6], such as those occurring during pipe bursts, segment isolation, or excessive water use, may cause service pressure to fall below the desired value. In these scenarios, DDA was noticed to yield poor prediction of nodal heads and pipe flows in the WDN. To solve this issue, the pressure-driven approach (PDA) was proposed in [7] and then considered by numerous authors, e.g., [8–22]. In PDA, the ratio of nodal outflow to nodal demand can be expressed as a function of a nodal pressure head by using one of the numerous formulations available in the scientific literature [7,9,12,13,17,20].

Whereas lots of research was dedicated to setting up increasingly robust and efficient algorithms for WDN solution (e.g., [15,16,21]), the effects of the various formulations proposed in terms of WDN solution results and algorithm convergence speed have been analyzed marginally. This paper aims to contribute to this issue.

In the following sections, first the methodology is described, including the pressure-driven formulations considered and the model used for WDN solution. The methodology is followed by the applications to two case studies and by the conclusions of the work.

2. Materials and Methods

In the present section, the pressure-driven formulations compared are presented (Section 2.1) followed by the algorithm used for WDN resolution (Section 2.2).

2.1. Pressure-Driven Formulations

The pressure-driven formulations considered in this work to express the actual nodal outflow q as a function of the actual nodal pressure head h and demand d are those of Wagner et al. [9], Fujiwara and Li [12], Tucciarelli et al. [13], Tanyimboh and Templeman with default parameters [17], and with calibration proposed by Ciaponi et al. [20]. All formulations consider the definition of a desired pressure head h_{des} for full demand satisfaction. The Wagner et al. [9], Fujiwara and Li [12], and Tanyimboh and Templeman with default parameters [17] formulations also consider a minimum pressure head h_{mim} to have outflow while the Tucciarelli et al. [13] and Ciaponi et al. [20] formulations were developed considering $h_{mim} = 0$.

In the formulations of Wagner et al. [9], Fujiwara and Li [12], and Tucciarelli et al. [13], $q/d = 0$ and $q/d = 1$ for $h \leq h_{min}$ and $h \geq h_{des}$, respectively.

For $h_{min} \leq h \leq h_{des}$, the Wagner et al. [9] formulation yields:

$$\frac{q}{d} = \left(\frac{h - h_{min}}{h_{des} - h_{min}} \right)^{1/\gamma}, \tag{1}$$

with coefficient γ usually set at 2 to re-obtain the same kind of relationship of the outflow through an orifice. Though being the most widely used, this formulation has the drawback of presenting derivative discontinuities for $h = h_{min}$ and $h = h_{des}$. This drawback is overcome in the formulations of Fujiwara and Li [12] and Tucciarelli et al. [13].

For $h_{min} \leq h \leq h_{des}$, the Fujiwara and Li [12] formulation yields:

$$\frac{q}{d} = \frac{(h - h_{min})^2 (3h_{des} - 2h - h_{min})}{(h_{des} - h_{min})^3}. \tag{2}$$

For $h_{min} \leq h \leq h_{des}$, the Tucciarelli et al. [13] formulation yields:

$$\frac{q}{d} = \sin^2 \left(\frac{\pi}{2} \frac{h}{h_{des}} \right). \tag{3}$$

Unlike the formulations of Wagner et al. [9], Fujiwara and Li [12], and Tucciarelli et al. [13], that of Tanyimboh and Templeman [17] is based on the following single smooth relationship, which holds for all of the values of h:

$$\frac{q}{d} = \frac{\exp(Y + Bh)}{1 + \exp(Y + Bh)}, \tag{4}$$

where, in absence of field data, the default values of the parameters Y and B are calculated through the following expressions:

$$Y = \frac{-4.595h_{des} - 6.907h_{min}}{h_{des} - h_{min}}, \tag{5}$$

$$B = \frac{11.502}{h_{des} - h_{min}}. \tag{6}$$

Using the same structure as (4), Ciaponi et al. [20] developed statistically expressions for Y and B with the objective to reproduce, on average, the complex and varied set of phenomena governing the actual water delivery at each network node. The resulting formulation, proposed for flat sites, is:

$$\frac{q}{d} = \frac{\exp(-3.178 + 8.214h/h_{des})}{1 + \exp(-3.178 + 8.214h/h_{des})}. \tag{7}$$

The graphical comparison of the five formulations presented above is reported in Figure 1 for $h_{min} = 0$ and $h_{des} = 20$, showing that the Fujiwara et al. [12] and Tucciarelli et al. [13] formulations give very similar values of q/d, which are lower than the other formulations over almost the entire range of h/h_{des}. The Wagner et al. [9] formulation gives the highest values of q/d up to $h/h_{des} = 0.5$, while being overcome by the Tanyimboh and Templeman [17] and Ciaponi et al. [20] formulations to the right of this value. As for the last two formulations, it must be remarked that they behave similarly over the entire range of h/h_{des} and give higher values of q/d to the right and to the left of $h/h_{des} = 0.5$, respectively. Both formulations yield $q/d \approx 1$ for $h/h_{des} = 1$ and a q/d higher than 0 for $h/h_{des} = 0$ (especially the latter). The explanation for these positive values is that, even if $h = h_{min} = 0$ enables no outflow at the nodal ground elevation, some outflow is still feasible from underground floors.

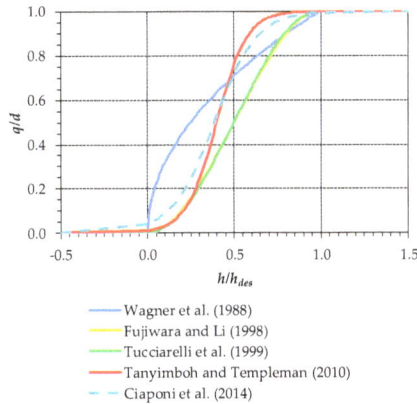

Figure 1. Ratio of outflow q to demand d as a function of the ratio of the generic pressure head h to the desired pressure head h_{des} for the various formulations.

2.2. WDN Resolution

WDN resolution consists in solving the momentum and continuity Equation (8) in vector form [3]. This enables us to obtain the unknown vectors **Q** and **H** of water discharges and heads for pipes and demanding nodes, respectively, starting from the vector **H$_0$** of heads at source nodes.

$$\begin{cases} \mathbf{A_{11}Q + A_{12}H = -A_{10}H_0} & \text{momentum equation} \\ \mathbf{A_{21}Q = q_{tot}} & \text{continuity equation} \end{cases} \tag{8}$$

In Equation (8), matrices **A$_{10}$** and **A$_{12}$** are obtained starting from the topological incidence matrix, **A**, by considering only its columns associated with the nodes with a fixed and unknown head, respectively. In fact, the number of rows and columns of matrix **A** is equal to the number of pipes and nodes, respectively. For the generic pipe, matrix **A** enables us to distinguish the upstream and downstream nodes (matrix elements equal to -1 and 1, respectively) from nodes not belonging to the pipe (matrix element equal to 0). **A$_{11}$** is a diagonal matrix with the number of rows and columns equal to the number of pipes. The generic diagonal element is defined as:

$$A_{11}(j,j) = r_j \left| Q_j \right|_{\alpha_j - 1}, \tag{9}$$

with r_j and α_j being the coefficient and the exponent, respectively, in the head loss equation relative to the j-th pipe.

In Equation (8), **q$_{tot}$** is the vector of nodal outflows obtained as the sum of the outflow **q** to the users and the vector **q$_l$** of leakage allocated to the demanding nodes. Whereas the generic element q of vector **q** is evaluated using whatever formulations presented in Section 2.1, the generic element q_l of $\mathbf{q_l}$ can be calculated through the following formula derived from [13]:

$$q_l = \sum_{i=1}^{ncp} \frac{CL_i}{2} h^n \tag{10}$$

where C and L_i are the WDN leakage coefficient and the length of the i-th of the *ncp* pipes connected to the generic node, respectively.

The global gradient algorithm (GGA) enables iterative resolution of Equation (8). This enables us to derive vectors **H** and **Q** at iteration $\tau + 1$ starting from those at iteration τ [3,23]:

$$\mathbf{H}^{\tau+1} = (\mathbf{G}^\tau)^{-1}\left\{\mathbf{A_{21}}(\mathbf{D_{11}^\tau})^{-1}[(\mathbf{D_{11}^\tau} - \mathbf{A_{11}^\tau})\mathbf{Q}^\tau - \mathbf{A_{10}H_0}] - \mathbf{q_{tot}}^\tau\right\}$$
$$\mathbf{Q}^{\tau+1} = \mathbf{Q}^\tau - (\mathbf{D_{11}^\tau})^{-1}(\mathbf{A_{11}^\tau}\mathbf{Q}^\tau + \mathbf{A_{12}H}^{\tau+1} + \mathbf{A_{10}H_0}) \tag{11}$$

where $\mathbf{D_{11}}^\tau$ and $\mathbf{A_{11}}^\tau$ are the values of $\mathbf{D_{11}}$ and $\mathbf{A_{11}}$ at iteration τ, where matrix $\mathbf{D_{11}}$ is a diagonal matrix, given by:

$$d\mathbf{A_{11}} = \mathbf{D_{11}}d\mathbf{QG} = \mathbf{A_{21}}(\mathbf{D_{11}})^{-1}\mathbf{A_{12}}. \tag{12}$$

At iteration $\tau = 0$, the elements of vector $\mathbf{Q^0}$ can be set at a small starting value, e.g., 0.001 m^3/s. Furthermore, $q_l = 0$ and $q = d$ at all nodes, entailing that $\mathbf{q_{tot}}^0$ is equal to **d** (vector of nodal demands).

In the following iterations, vector $\mathbf{q_{tot}^\tau}$ is a function of **d**, which is independent from **H**, and of head **H**. Vector $\mathbf{q_{tot}^\tau}$ in Equation (11) could be set at $\mathbf{q_{tot}}(\mathbf{d}, \mathbf{H}^{\tau-1})$, which is the vector of total nodal outflows, obtained as the sum of **q** and $\mathbf{q_l}$, evaluated though whatever formulation in Section 2.1 and through Equation (10), respectively, while considering the vector **H** of heads at the previous iteration $\tau - 1$. However, this was noticed to yield numerical oscillations by various authors [15,19,21]. To tackle this problem, the following underrelaxation method can be used to speed up the convergence of the GGA:

$$\mathbf{q_{tot}^\tau} = \mathbf{q_{tot}^{\tau-1}} + \Omega^\tau\left[\mathbf{q_{tot}}\left(\mathbf{d}, \mathbf{H}^{\tau-1}\right) - \mathbf{q_{tot}^{\tau-1}}\right] \tag{13}$$

in which Ω^τ lying between 0 and 1 is an underrelaxation factor, which is set at 1 at $\tau = 1$. Then, across the iterations, it is reduced in the case of the occurrence of oscillations in **H**. To this end, at the generic iteration τ, it is calculated as:

$$\Omega^\tau = \begin{cases} \Omega^{\tau-1} & \text{if } \mathrm{norm}(\mathbf{H}^\tau - \mathbf{H}^{\tau-1}) < \mathrm{norm}(\mathbf{H}^{\tau-1} - \mathbf{H}^{\tau-2}) \\ \alpha\,\Omega^{\tau-1} & \text{if } \mathrm{norm}(\mathbf{H}^\tau - \mathbf{H}^{\tau-1}) \geq \mathrm{norm}(\mathbf{H}^{\tau-1} - \mathbf{H}^{\tau-2}) \end{cases} . \tag{14}$$

In Equation (14), α is set at a positive number lower than 1 (e.g., 0.5), to enable Ω^τ reduction.

Various stopping criteria can be used in this model, such as $\max|\mathbf{H}^\tau - \mathbf{H}^{\tau-1}| \leq T$, with T being a threshold, which can be set at the reasonable value of 0.001 m. The underrelaxation method is different from those used in [15,21], in that the underrelaxation is applied on nodal outflows rather than on nodal heads and pipe flows. Incidentally, the method presented above to prevent numerical oscillation is a refinement of that originally proposed in [14]. The refinement consists in the introduction of norm function as a criterion to vary Ω.

3. Application

3.1. Case Studies

The application of this work concerned two case studies (Figure 2a,b, respectively). The first is the skeletonized WDN of Santa Maria di Licodia, a small town in Sicily, Italy [24,25]. The network layout is made up of 34 nodes (of which 32 are with unknown head and 2 source nodes are with fixed head, i.e., nodes 33 and 34) and of 41 pipes. The features of the WDN nodes and pipes are reported in the referenced work. At the generic demanding node, h_{min} and h_{des} were set at 0 and 20 m, respectively.

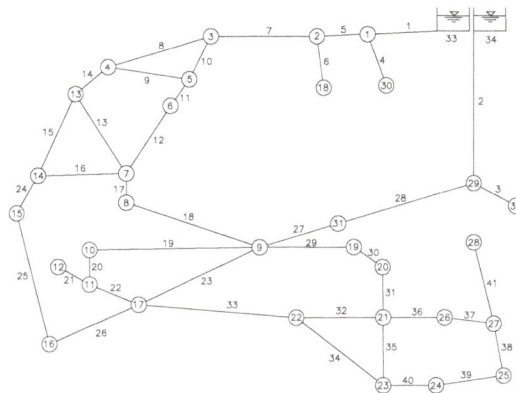

(a)

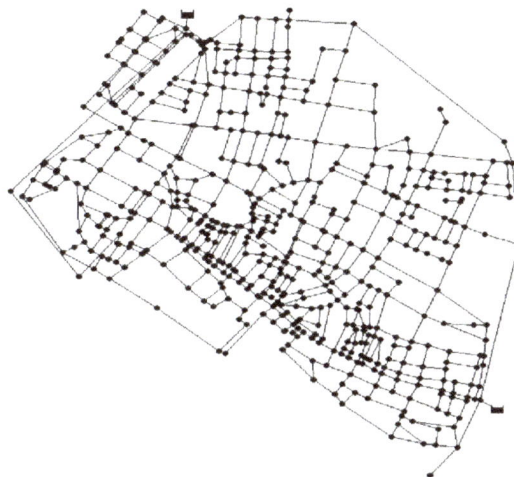

(b)

Figure 2. Water distribution networks (WDNs) for the two case studies (**a,b**). In the first case study with source nodes 33 and 34, pipe numbers are close to the pipes and the numbers of demanding nodes are inside circles.

The two source nodes, i.e., nodes 33 and 34, have a ground elevation of 477.5 m above sea level and a mean daily head of 480.77 m. The topography of the network features declining nodal ground levels as the distance from the source increases. The ground elevation of demanding nodes varies between about 395 and 465 m above sea level.

The whole network demand is 18.5 L/s.

As to leakage, values of the coefficient $C_L = 2.79 \times 10^{-8}$ m$^{0.82}$/s and of the exponent $n_{leak} = 1.18$ were assumed for all of the pipes of the network. The values of C_L and n_{leak} lead to a daily leakage volume of about 1250 m^3, which is about 44% of the water volume leaving the source nodes.

The second case study of the paper is made up of a real WDN in Northern Italy featuring 536 nodes with unknown head, 825 pipes, and 2 reservoirs (Figure 2b) [26]. At the generic demanding node, h_{min} and h_{des} were set at 0 and 23 m, respectively. The head of the two reservoirs is 30 m above sea level.

Unlike the first network, the second network has flat topography with all nodal ground levels set at 0 m above sea level. The whole network demand is 367 L/s.

As to leakage, values of the coefficient $C_L = 1.99 \times 10^{-7}$ m$^{2.5}$/s and of the exponent $n_{leak} = 0.5$ were assumed for all of the pipes of the WDN. The values of C_L and n_{leak} lead to a daily leakage volume of about 7920 m^3, which is about 20% of the water volume leaving the source nodes.

The two WDNs were solved in snapshot simulations using the five pressure-driven formulations described in Section 2.1. In these simulations, nodal demands were varied considering a uniform multiplying coefficient, α_{dem}, ranging from 1 to 20 with step 1. This resulted in a total framework of 100 simulations (5 pressure driven formulations × 20 demand increase steps) for each WDN. The objective of the simulations was to investigate how differently the pressure-driven formulations react to changes in network operating conditions. In fact, the progressively increasing demand puts the WDNs to an increasingly severe test by lowering nodal pressure heads below h_{des} and, as a result, by triggering the pressure-driven dependence $q(h)$ (Figure 1). In this analysis, the Wagner et al. [9] formulation was taken as benchmark.

3.2. Results

As for the first case study, the first analysis concerns the relationship between the total number k of iterations and α_{dem} (see graph in Figure 3). This graph shows $k(\alpha_{dem})$ for all of the formulations. For the formulation of Wagner et al. [9], which is the most commonly used in the scientific literature, WDN resolution tends to be slower and slower when α_{dem} grows up to $\alpha_{dem} = 20$.

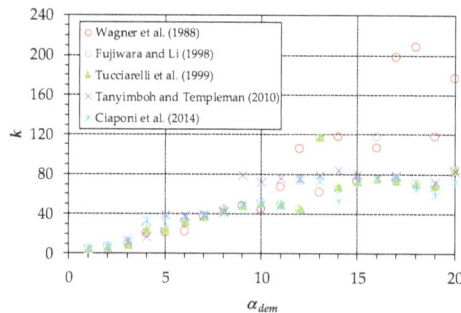

Figure 3. First case study. Number k of iterations as a function of α_{dem} for the various formulations.

The other formulations behave similarly to Wagner et al. [9] up to $\alpha_{dem} = 10$. Above this value, they tend to converge faster, that is with lower values of k. This seems to suggest that the other formulations suffer from the increase in α_{dem} less than the Wagner et al. [9] formulation.

The graphs in Figures 4 and 5 and Table 1 report results concerning analyses and comparisons in terms of outflow q to users. Those in Figure 4 compare the values of q obtained with the Wagner et al. [9] formulation with those obtained with the four other formulations. For $\alpha_{dem} = 1$, the values are very close because the pressure-driven dependence of q has not been triggered. For larger values of α_{dem}, some differences arise. Far from the 0, the four other formulations yield close values to those of the Wagner et al. [9] formulation, with differences of a similar order of magnitude. However, the differences obtained through the Fujiwara and Li [12] and Tucciarelli et al. [13] formulations are always negative (underestimation). Those obtained through the Tanyimboh and Templeman [17] and of Ciaponi et al. [20] formulations, instead, are alternatively positive and negative. Close to the 0 of the Wagner et al. [9] formulation, all the other formulations tend to yield larger values of q. This behavior may be due to the softer declining trend that the other formulations have in Figure 1 when h/h_{des} goes down to 0.

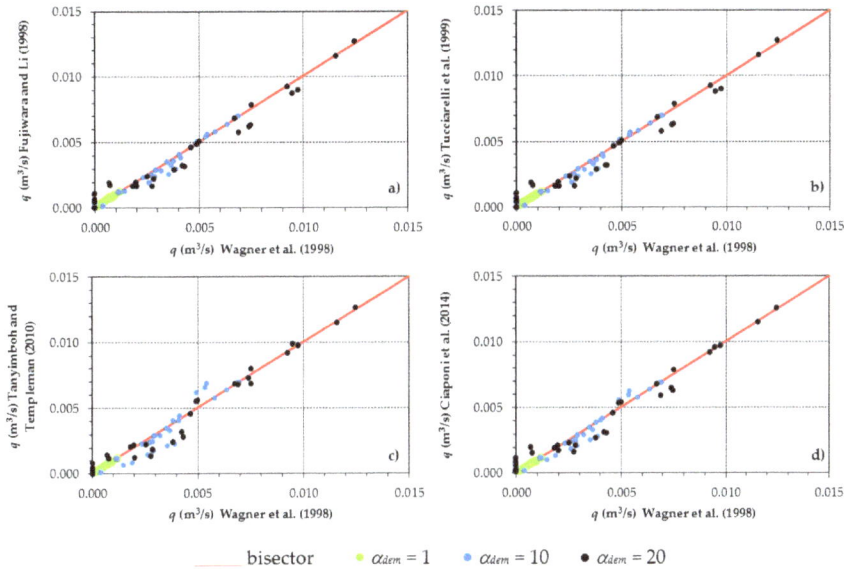

Figure 4. First case study. Comparison of the various formulations, (**a**) Fujiwara and Li [12], (**b**) Tucciarelli et al. [13], (**c**) Tanyimboh and Templeman [17] and (**d**) Ciaponi et al. [20], in terms of outflow q to users.

A quantitative estimation of the comparison shown in Figure 4 is reported in the following Table 1. This table points out that the mean absolute deviation in q of the various formulations compared to the Wagner et al. [9] formulation tends to grow with α_{dem}. Overall, the formulations yield almost no deviation for $\alpha_{dem} = 1$, in correspondence to which the pressure-driven behavior is not fully activated. Indeed, it concerns only a few demanding nodes. For $\alpha_{dem} = 10$ and $\alpha_{dem} = 20$, they behave similarly, with the Tanyimboh and Templeman [17] and Ciaponi et al. [20] formulations providing the largest deviations, respectively.

Table 1. First case study. Mean absolute deviation (m^3/s) of the various formulations in terms of outflow q to users compared to the Wagner et al. [9] formulation for the three values of α_{dem}.

Formulation	$\alpha_{dem} = 1$	$\alpha_{dem} = 10$	$\alpha_{dem} = 20$
Fujiwara and Li [12]	0	0.00022	0.00049
Tucciarelli et al. [13]	0	0.00024	0.00049
Tanyimboh and Templeman [17]	0	0.00042	0.00044
Ciaponi et al. [20]	0	0.00029	0.00051

Figure 5 shows the trend of the global demand satisfaction rate $\sum q / \sum d$, as a function of α_{dem}, pointing out that all of the formulations behave similarly. This happens because the overestimations and underestimations compared to the Wagner et al. [9] formulations tend to cancel each other out. As an example of the results, for $\alpha_{dem} = 10$, the formulations of Fujiwara and Li [12] and of Tucciarelli et al. [13] give a satisfaction rate of 0.52. The Wagner et al. [9] formulation gives a satisfaction rate of 0.55. Finally, the formulations of Tanyimboh and Templeman [17] and of Ciaponi et al. [20] give an intermediate value of 0.53.

The graphs in Figure 6 compare the nodal pressure heads h obtained with the Fujiwara et al. [12], Tucciarelli et al. [13], Tanyimboh and Templeman [17], and Ciaponi et al. [20] formulations with those obtained with the Wagner et al. [9] formulation. In all cases, the dots are well-aligned along the bisector, apart from close to the 0 of the Wagner et al. formulation, in correspondence to which the other formulations seem to overestimate *h*.

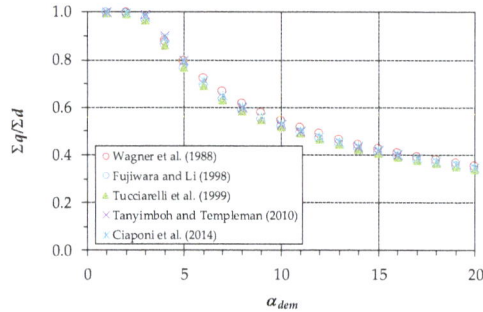

Figure 5. First case study. Satisfaction rate $\sum q / \sum d$ for the various formulations as a function of α_{dem}.

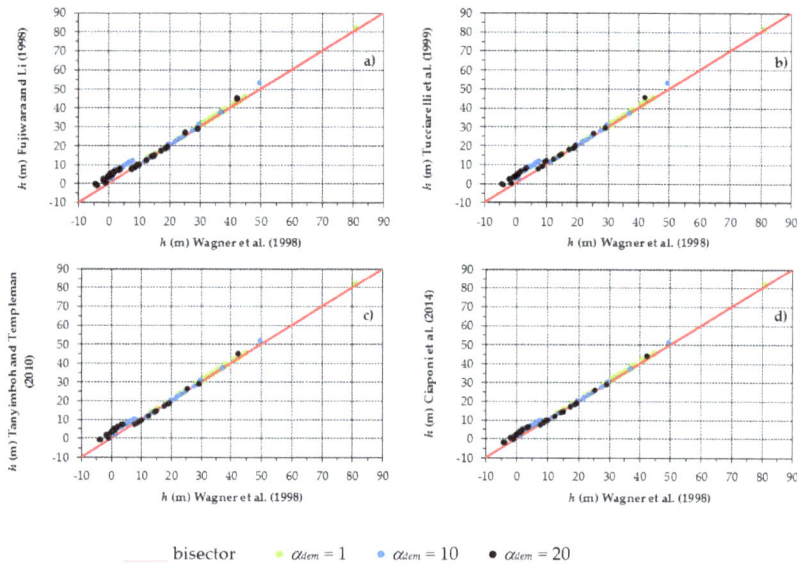

Figure 6. First case study. Comparison of the various formulations, (**a**) Fujiwara and Li [12], (**b**) Tucciarelli et al. [13], (**c**) Tanyimboh and Templeman [17] and (**d**) Ciaponi et al. [20], in terms of nodal pressure head *h*.

A quantitative estimation of the comparison shown in Figure 6 is reported in the following Table 2. This table points out that the mean absolute pressure head deviation of the various formulations compared to the Wagner et al. [9] formulation tends to grow with α_{dem}. Overall, the Ciaponi et al. [20] and the Tucciarelli et al. [13] formulations are the closest and furthest, respectively.

Figure 7 shows the generalized resilience/failure index (GRF) [27] related to the pressure head distribution in the WDN. This index, ranging from −1 to 1, represents an estimate of WDN resilience through the pressure-driven approach when leakage is present. The positive and negative values of GRF are associated with WDNs under power surplus and deficit conditions, respectively. The graph in Figure 7 shows that, despite the differences in *h* remarked in Figure 6, the formulations yield similar decreasing trends of GRF, as was the case with the satisfaction rate in Figure 5.

Table 2. First case study. Mean absolute deviation (m) of the various formulations in terms of nodal pressure head h compared to the Wagner et al. [9] formulation for the three values of α_{dem}.

Formulation	$\alpha_{dem} = 1$	$\alpha_{dem} = 10$	$\alpha_{dem} = 20$
Fujiwara and Li [12]	0.0012	2.65	2.69
Tucciarelli et al. [13]	0.0010	2.68	2.76
Tanyimboh and Templeman [17]	0.0024	1.84	2.26
Ciaponi et al. [20]	0.0013	1.54	1.54

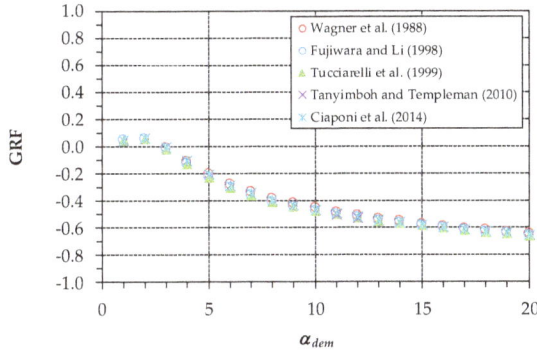

Figure 7. First case study. Generalized resilience/failure index (GRF) for the various formulations as a function of α_{dem}.

The similar values of GRF are because this index depends on both h and q. In fact, the variations in these variables across the various formulations tend to cancel each other out.

The results of the applications to the second case study are reported in the following Figures 8–12. These figures are homologous to Figures 3–7, respectively, of the first case study. The comparison of Figures 3 and 8 highlights that k tends to grow as α_{dem} grows also for the second case study. However, the beneficial effects of the Fujiwara et al. [12], Tucciarelli et al. [13], Tanyimboh and Templeman [17], and Ciaponi et al. [20] formulations in terms of k reduction, in comparison with the Wagner et al. [9] formulation, are attenuated in the second case study. This may be because of the more interlinked structure of the second network, which helps to regularize pipe flows and nodal heads across the iterations in WDN resolution. All of the other figures have similar interpretations to the homologous figures of the first case study.

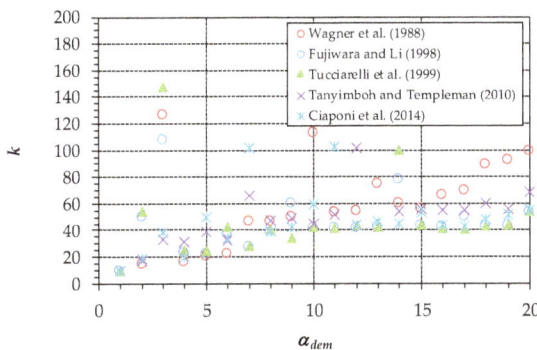

Figure 8. Second case study. Number k of iterations as a function of α_{dem} for the various formulations.

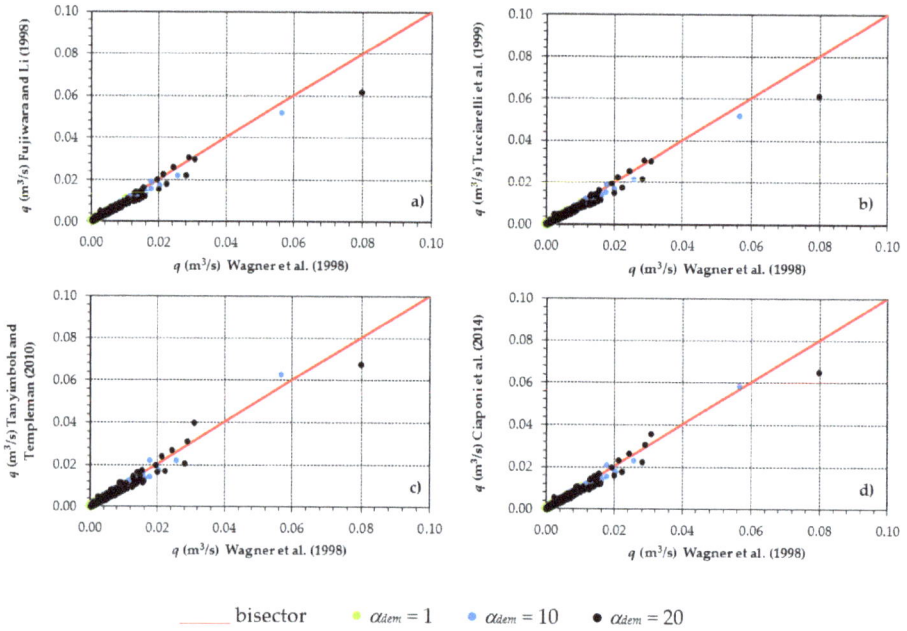

Figure 9. Second case study. Comparison of the various formulations, (**a**) Fujiwara and Li [12], (**b**) Tucciarelli et al. [13], (**c**) Tanyimboh and Templeman [17] and (**d**) Ciaponi et al. [20], in terms of outflow *q* to users.

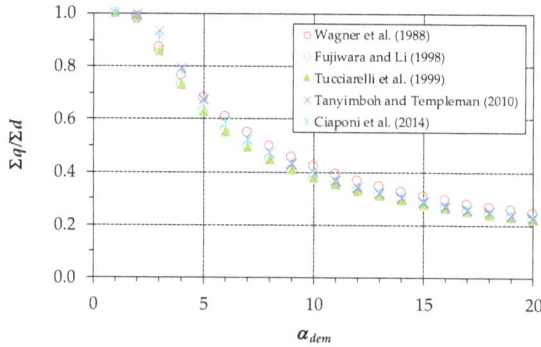

Figure 10. Second case study. Satisfaction rate $\sum q/\sum d$ for the various formulations as a function of α_{dem}.

The fact that the *h* overestimation of the Fujiwara et al. [12], Tucciarelli et al. [13], Tanyimboh and Templeman [17], and Ciaponi et al. [20] formulations compared to the Wagner et al. [9] formulation seems to be larger in the second case study (compare Figure 11 with Figure 6) is mainly due to the scale used in the graph (*h* values ranging from 0 to 30 m in the second case study versus *h* values ranging from −5 to above 80 m in the first case study). This is clear from Figure 13, where Figure 6 of the first case study is plotted with similar scale to Figure 11.

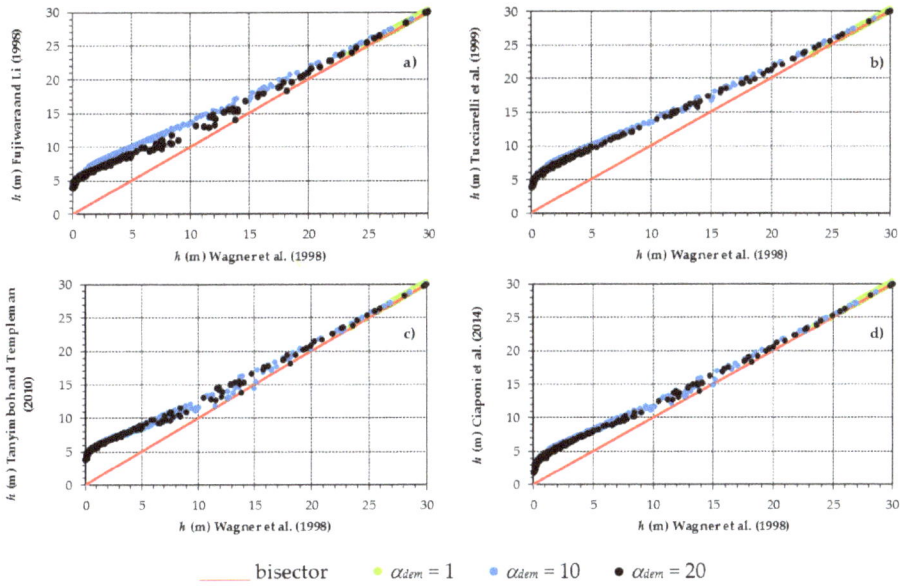

Figure 11. Second case study. Comparison of the various formulations, (**a**) Fujiwara and Li [12], (**b**) Tucciarelli et al. [13], (**c**) Tanyimboh and Templeman [17] and (**d**) Ciaponi et al. [20], in terms of nodal pressure head h.

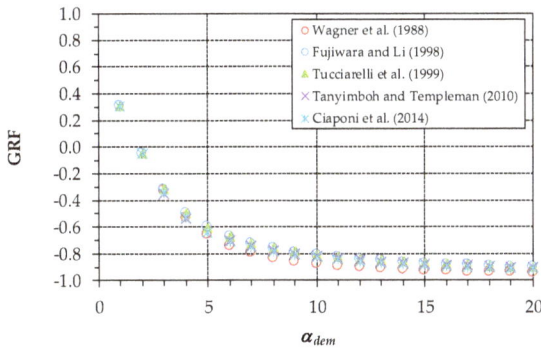

Figure 12. Second case study. GRF for the various formulations as a function of α_{dem}.

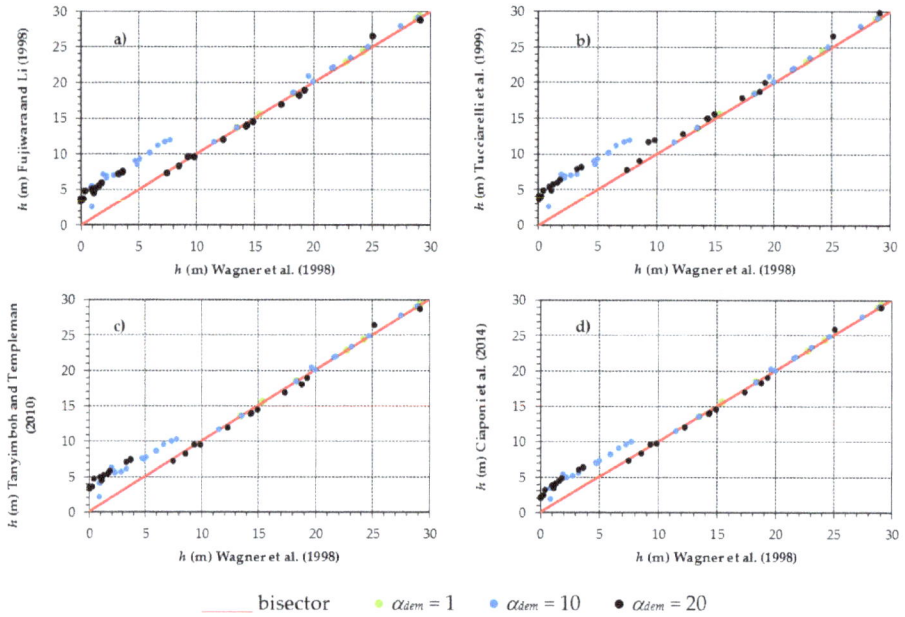

Figure 13. First case study. Comparison of the various formulations, (**a**) Fujiwara and Li [12], (**b**) Tucciarelli et al. [13], (**c**) Tanymboh and Templeman [17] and (**d**) Ciaponi et al. [20], in terms of nodal pressure head *h*. Zoom of Figure 6 with scale similar to Figure 11.

4. Conclusions

In this paper, a comparison of various pressure-driven formulations, aimed at expressing nodal outflows to users as a function of nodal demands and of the actual nodal pressure, was presented. To this end, the widely adopted formulation of Wagner et al. [9] was taken as benchmark. The comparison was made in the framework of the snapshot steady flow simulation of two different WDNs. These WDNs represent the main skeleton of a poorly interconnected WDN situated in a hilly territory and a very interconnected real WDN situated in a flat territory, respectively. Nodal demands were largely amplified to stress the pressure-driven behavior of the WDNs.

The main outcomes of the work are:

- The increase in nodal demands tends to cause the growth of the number of iterations for WDN algorithm convergence due to the activation of the pressure-driven relationship.
- Especially in the poorly interconnected WDN, the Fujiwara and Li [12], Tucciarelli et al. [13], Tanymboh and Templeman [17], and Ciaponi et al. [20] formulations yield faster convergence for large nodal demands compared to the Wagner et al. [9] formulation.
- All formulations similarly simulate nodal outflows.
- The Fujiwara and Li [12], Tucciarelli et al. [13], Tanymboh and Templeman [17], and Ciaponi et al. [20] formulations overestimate nodal pressure heads compared to the Wagner et al. [9] formulation.

Though the effects of the various formulations were compared, it must be noted that:

- The different effects of the formulations were accentuated in this work due to the large and uniform amplification of WDN demands. In real cases, the pressure-driven behavior is usually remarkable in small pressure-deficient areas, due to hydrant activations or segment isolations. Therefore, the differences are expected to be more confined.

- Only the comparison with experimental data can reveal which formula is the most consistent with the real WDN behavior.

- In the absence of experimental data, the use of formulations based on statistical simulation of a varied set of situations and phenomena governing the actual water delivery, such as the formulation of Ciaponi et al. [20], may be preferable.

As for the last remarks, it must be noted that parameterizing pressure-driven formulations on the field may be a heavy task. In fact, at a generic WDN node, the outflow q is the only variable that can be measured through metering devices. Demand d can only be supposed. Another tricky factor lies in the fact that, as stated above, the pressure-driven behavior of the WDN is often temporary and unpredictable. Therefore, the only viable option in most circumstances may be the parameterization of the formulations based on statistical simulations of outflow, as was done by Ciaponi et al. [20].

Author Contributions: Both authors contributed equally to the paper.

Conflicts of Interest: The authors declare no conflict of interest.

References

1. Wood, D.J.; Charles, O.A. Hydraulic network analysis using linear theory. *J. Hydraul. Division* **1970**, *96*, 1221–1234.
2. Isaacs, L.T.; Mills, K.G. Linear theory methods for pipe network analysis. *J. Hydraul. Division* **1980**, *106*, 1191–1201.
3. Todini, E.; Pilati, S. *A Gradient Algorithm for the Analysis of Pipe Networks*; Wiley: London, UK, 1988; pp. 1–20.
4. Creaco, E.; Franchini, M. Comparison of Newton-Raphson Global and Loop Algorithms for Water Distribution Network Resolution. *J. Hydraul. Eng.* **2014**, *140*, 313–321. [CrossRef]
5. Walski, M.; Chase, D.; Savic, D.; Grayman, W.; Beckwith, S.; Koelle, E. *Advanced Water Distribution Modelling and Management*; Haestad: Waterbury, CT, USA, 2003.
6. Wu, Z.Y.; Walski, T. Pressure dependent hydraulic modelling for water distribution systems under abnormal conditions. In Proceedings of the IWA World Water Congress and Exhibition, Beijing, China, 10–14 September 2006.
7. Bhave, P.R. Node flow analysis of water distribution systems. *J. Transp. Eng.* **1981**, *107*, 457–467.
8. Germanopoulos, G. A technical note on the inclusion of pressure dependent demand and leakage terms in water supply network models. *J. Civ. Eng. Syst.* **1985**, *2*, 171–179. [CrossRef]
9. Wagner, B.J.M.; Shamir, U.; Marks, D.H. Water distribution reliability: Simulation method. *J. Water Resour. Plan. Manag.* **1988**, 276–294. [CrossRef]
10. Chandapillai, J. Realistic simulation of water distribution system. *J. Transp. Eng.* **1991**, 258–263. [CrossRef]
11. Gupta, R.; Bhave, P.R. Comparison of methods for predicting deficient-network performance. *J. Water Resour. Plan. Manag.* **1996**, 214–217. [CrossRef]
12. Fujiwara, O.; Li, J. Reliability analysis of water distribution networks in consideration of equity, redistribution, and pressuredependent demand. *J. Water Resour. Res.* **1998**, *34*, 1843–1850. [CrossRef]
13. Tucciarelli, T.; Criminisi, A.; Termini, D. Leak Analysis in Pipeline System by Means of Optimal Value Regulation. *J. Hydraul. Eng.* **1999**, *125*, 277–285. [CrossRef]
14. Alvisi, S.; Franchini, M. Near-optimal rehabilitation scheduling of water distribution systems based on a multi-objective genetic algorithm. *Civ. Eng. Environ. Syst.* **2006**, *23*, 143–160. [CrossRef]
15. Giustolisi, O.; Savic, D.; Kapelan, Z. Pressure-driven demand and leakage simulation for water distribution networks. *J. Hydraul. Eng.* **2008**, 626–635. [CrossRef]
16. Wu, Z.Y.; Wang, R.H.; Walski, T.M.; Yang, S.Y.; Bowdler, D.; Baggett, C.C. Extended global-gradient algorithm for pressure-dependent water distribution analysis. *J. Water Resour. Plan. Manag.* **2009**. [CrossRef]
17. Tanyimboh, T.; Templeman, A. Seamless pressure-deficient water distribution system model. *J. Water Manag.* **2010**, *163*, 389–396. [CrossRef]
18. Creaco, E.; Franchini, M.; Alvisi, S. Evaluating water demand shortfalls in segment analysis. *Water Resour. Manag.* **2012**, *26*, 2301–2321. [CrossRef]

19. Siew, C.; Tanyimboh, T.T. Pressure-dependent EPANET extension. *J. Water Resour. Manag.* **2012**, *26*, 1477–1498. [CrossRef]
20. Ciaponi, C.; Franchioli, L.; Murari, E.; Papiri, S. Procedure for defining a pressure-outflow relationship regarding indoor demands in pressure-driven analysis of water distribution networks. *Water Resour. Manag.* **2015**, *29*, 817–832. [CrossRef]
21. Elhay, S.; Piller, O.; Deuerlein, J.; Simpson, A. A robust, rapidly convergent method that solves the water distribution equations for pressure-dependent models. *J. Water Resour. Plan. Manag.* **2015**, 04015047. [CrossRef]
22. Pacchin, E.; Alvisi, S.; Franchini, M. Analysis of Non-Iterative Methods and Proposal of a New One for Pressure-Driven Snapshot Simulations with EPANET. *Water Resour. Manag.* **2017**, *31*, 75–91. [CrossRef]
23. Todini, E.; Rossman, L.A. Unified framework for deriving simultaneous equation algorithms for water distribution networks. *J. Hydraul. Eng.* **2013**, *139*, 511–526. [CrossRef]
24. Pezzinga, G. Procedure per la riduzione delle perdite mediante il controllo delle pressioni. In *Ricerca e Controllo delle Perdite Nelle Reti di Condotte. Manuale per una Moderna Gestione degli Acquedotti*; Brunone, B., Ferrante, M., Meniconi, S., Eds.; CittàStudiEdizioni: Novara, Italy, 2008. (In Italian)
25. Creaco, E.; Pezzinga, G. Embedding Linear Programming in Multi Objective Genetic Algorithms for Reducing the Size of the Search Space with Application to Leakage Minimization in Water Distribution Networks. *Environ. Model. Softw.* **2015**, *69*, 308–318. [CrossRef]
26. Creaco, E.; Franchini, M. Fast network multi-objective design algorithm combined with an a posteriori procedure for reliability evaluation under various operational scenarios. *Urban Water J.* **2012**, *9*, 385–399. [CrossRef]
27. Creaco, E.; Franchini, M.; Todini, E. Generalized Resilience and Failure Indices for Use with Pressure-Driven Modeling and Leakage. *J. Water Resour. Plan. Manag.* **2017**, *142*, 04016019. [CrossRef]

water MDPI

Article

Comparison of Algorithms for the Optimal Location of Control Valves for Leakage Reduction in WDNs

Enrico Creaco [1] and Giuseppe Pezzinga [2,*]

[1] Dipartimento di Ingegneria Civile e Architettura, Università di Pavia, via Ferrata 3,
    27100 Pavia, Italy; creaco@unipv.it
[2] Dipartimento di Ingegneria Civile e Architettura, Università di Catania, via Santa Sofia 64,
    95123 Catania, Italy
* Correspondence: giuseppe.pezzinga@unict.it

Received: 20 February 2018; Accepted: 10 April 2018; Published: 12 April 2018

Abstract: The paper presents the comparison of two different algorithms for the optimal location of control valves for leakage reduction in water distribution networks (WDNs). The former is based on the sequential addition (SA) of control valves. At the generic step N_{val} of SA, the search for the optimal combination of N_{val} valves is carried out, while containing the optimal combination of $N_{val} - 1$ valves found at the previous step. Therefore, only one new valve location is searched for at each step of SA, among all the remaining available locations. The latter algorithm consists of a multi-objective genetic algorithm (GA), in which valve locations are encoded inside individual genes. For the sake of consistency, the same embedded algorithm, based on iterated linear programming (LP), was used inside SA and GA, to search for the optimal valve settings at various time slots in the day. The results of applications to two WDNs show that SA and GA yield identical results for small values of N_{val}. When this number grows, the limitations of SA, related to its reduced exploration of the research space, emerge. In fact, for higher values of N_{val}, SA tends to produce less beneficial valve locations in terms of leakage abatement. However, the smaller computation time of SA may make this algorithm preferable in the case of large WDNs, for which the application of GA would be overly burdensome.

Keywords: valve; pressure; leakage; optimization

---

1. Introduction

The pursuit of benefits in terms of leakage and pipe burst reduction as well as of infrastructure life extension has spurred water utilities to perform the active control of service pressure in water distribution networks (WDNs) [1–3]. In fact, if service pressure is reduced without violating pressure constraints for water supply, positive effects arise without the service effectiveness being affected. The active control of service pressure can also be performed through devices that enables electric power generation [4,5].

When WDN sources have pressure surplus compared to demanding nodes, the active control can be carried out by inserting control valves in key locations of the network. Two main lines of research are currently underway in this context: the former, which started with the paper by Jowitt and Xu [6], concerns the optimal location and control of valves, while the latter, which started with the paper by Campisano et al. [7], concerns the real time regulation of these devices.

In the former line of research, which is the topic of the present paper, some initial works were dedicated to setting up algorithms for the optimization of control valve settings to minimize daily leakage, while meeting minimum pressure requirements at WDN demanding nodes. In this context, Jowitt and Xu [6] proposed an algorithm based on iterated linear programming (LP) to minimize leakage at each time slot of the day. Vairavamoorthy and Lumbers [8], instead, proposed the use of sequential quadratic programming, with an objective function that allows minor violations in the targeted pressure requirements. Other papers [9–14] addressed both the issues of valve location and control. Reis et al. [9] tackled the optimization of valve locations and settings by using genetic algorithms (GA) and LP, respectively. In detail, LP was embedded in the genetic algorithm to search for the optimal valve settings for each location of control valves proposed as a solution by the genetic algorithm. Araujo et al. [10] and Cozzolino et al. [13] made use of genetic algorithms for both issues. Liberatore and Sechi [11] proposed a two-step procedure for the optimal valve location and control. In the former step, candidate sets for the location of valves are restricted to pipes defined based on hydraulic analysis. The meta-heuristic Scatter Search routines were used in the latter step to identify the best solution in the location and control problems by optimizing a weighted multi-objective function that considers the cost of inserting valves and the penalty for node pressures that do not meet the requirements. Ali [12] made exclusive use of genetic algorithms, in which he incorporated the physical knowledge of the WDN to improve the algorithm efficiency. De Paola et al. [14] made use of the harmony search approach to optimize both control valve locations and settings.

Whereas all the algorithms cited above are based on the single-objective approach, other algorithms [15–18] were conceived using the multi-objective approach, to construct Pareto fronts of optimal solutions in the tradeoff between number of control valves, as a surrogate for the installation cost, and daily leakage. Pezzinga and Gueli [15] proposed, for optimal valve location, a fully deterministic procedure, based on the sequential addition (SA) of beneficial valves up to a maximum number of valves installable in the WDN has been fixed. LP is used, instead, for optimal valve control. Nicolini and Zovatto [16] used a multi-objective GA to optimize both valve locations and control settings. Creaco and Pezzinga [17,18] used a GA to search for the optimal valve locations, and LP to optimize the control valve settings for each solution proposed by the GA.

Despite the plethora of works available on the subject, there are only a few comparative analyses. In fact, authors have often proposed new algorithms without testing them against those already available in the scientific literature. Furthermore, the merits and demerits of deterministic and probabilistic multi-objective approaches have never been compared. The lack of this kind of comparison in the scientific literature has motivated the present work. The algorithms of Pezzinga and Gueli [15] and of Creaco and Pezzinga [17,18] were chosen as representative for the comparison. The former was chosen because of its fully deterministic approach, which confers significant computational lightness on the methodology. The choice of the latter is due, instead, to its computational effectiveness, which was acknowledged [19] during the battle of background leakage assessment for water networks (BBLAWN) [20].

In the remainder of the paper, algorithms are first described and then applied to two WDNs.

2. Materials and Methods

In the present section, first the LP used in both SA [15] and GA [17,18] for the optimization of control valve settings is described (Section 2.1). Sections 2.2 and 2.3 are dedicated to describing the different approaches used in SA and GA, respectively, to tackle optimal valve locations.

2.1. Fitness Evaluation of the Generic Location of Control Valves

The generic location of control valves, proposed as solution by whatever algorithm (e.g., the algorithms described in Sections 2.2 and 2.3), can be evaluated in terms of installation cost and daily leakage volume W_L. The former assessment comes straight away after the control valves have been installed in selected pipes of the WDN model. In the applications of this work, the total number of control valves is considered as a surrogate for the cost. The latter assessment requires determination of the daily pattern of optimal valve settings in the context of WDN extended period simulation (EPS).

Let us assume a WDN with n_p pipes and n_n nodes = n demanding nodes + n_0 sources. WDN operation can be represented through a succession of $N_{\Delta t}$ time slots, all featuring the same duration Δt. At each time slot, in which source heads and nodal demands are assigned, pipe water discharges and nodal heads can be derived by solving the following equations, derived from previous works [17,18,21]:

$$\begin{cases} \mathbf{A_{11}MQ} + \mathbf{A_{12}H} = -\mathbf{A_{10}H_0} & \text{momentum equation} \\ \mathbf{A_{21}Q} = \mathbf{d} + \mathbf{q_L} & \text{continuity equation} \end{cases}, \tag{1}$$

where $\mathbf{H}$ ($n \times 1$) and $\mathbf{Q}$ ($n_p \times 1$) are the unknown vectors of nodal heads at the demanding nodes and of pipe water discharges, respectively. $\mathbf{H_0}$ ($n_0 \times 1$) and $\mathbf{d}$ ($n \times 1$) are the assigned vectors of source heads and nodal demands, respectively. Matrixes $\mathbf{A_{12}}$ and $\mathbf{A_{10}}$ come from the incidence topological matrix $\mathbf{A}$, with size $n_p \times n_n$. In the generic row of $\mathbf{A}$, associated with the generic network pipe, the generic element can take on the values 0, -1 or 1, whether the node corresponding to the matrix element is not at the end of the pipe, it is the initial node of the pipe, or it is the final node of the pipe, respectively. Starting from $\mathbf{A}$, matrix $\mathbf{A_{12}}$ ($n_p \times n$) is obtained by considering the columns corresponding to the n network nodes with unknown head. $\mathbf{A_{21}}$ ($n \times n_p$) is the transpose matrix of $\mathbf{A_{12}}$. Matrix $\mathbf{A_{10}}$ ($n_p \times n_0$) is obtained by considering the columns corresponding to the n_0 nodes with fixed head. $\mathbf{A_{11}}$ ($n_p \times n_p$) is a diagonal matrix, the elements of which identify the resistances of the n_p network pipes through the following relationship:

$$A_{11}(i, i) = \frac{b_i |Q_i|^{\alpha-1} L_i}{k_i^\gamma D_i^\beta}, \tag{2}$$

where D_i is the diameter of the i-th pipe and where roughness coefficient k_i, coefficient b_i and exponents α, β and γ depend on the formula used to express pipe head losses. As an example, $\alpha = 2$, $\beta = 5.33$ and $\gamma = 2$ in the Strickler formula. Matrix $\mathbf{M}$ in Equation (1) is used to increase the resistance of the N_{val} pipes fitted with the control valve. It is a diagonal matrix ($n_p \times n_p$), in which the diagonal elements corresponding to the N_{val} pipes fitted with control valve are equal to V_k ($k = 1, \ldots, N_{val}$), whereas those corresponding to the n_p-N_{val} pipes without valve are equal to 1. Indeed, V_k is the valve setting ranging from 0 to 1. For the generic pipe subject to a certain head loss, it represents the ratio of the pipe water discharge in the presence of the regulated valve to the water discharge in its absence. The extreme values of the range represent the fully closed and fully open valve, respectively.

In Equation (1), the vector $\mathbf{q_L}$ ($n \times 1$) of leakage allocated to network demanding nodes can be calculated starting from the following Equation (3):

$$\mathbf{q_L} = \frac{|\mathbf{A_{21}}| \mathbf{Q_L}}{2}, \tag{3}$$

where the elements of the vector $\mathbf{Q_L}$ ($n_p \times 1$) of leakage outflows from WDN pipes can be assessed through the following relationship [6]:

$$Q_{Li} = C_{L,i} L_i h_i^{n_{leak}}, \tag{4}$$

where, for the i-th pipe, h_i and L_i are the mean pressure head (ratio of pressure to specific weight of water) and the length, respectively. $C_{L,i}$ and n_{leak} are empirical coefficients. The mean pressure head h_i can be calculated as:

$$h_i = \frac{H_{i,1} + H_{i,2} - z_{i,1} - z_{i,2}}{2}, \tag{5}$$

where $H_{i,1}$, $H_{i,2}$ and $z_{i,1}$, $z_{i,2}$ are the heads and elevations, respectively, for the end nodes of the pipe.

At each time slot of WDN operation, the vector $\mathbf{V}$ of valve settings V_k ($k = 1, \ldots, N_v$) can be optimized to minimize the total leakage volume $W_{L,j}$ from the network, while meeting the minimum pressure head requirement h_{des} at the demanding nodes:

$$W_{L,j} = \sum_{i=1}^{n_p} Q_{Li} \Delta t. \tag{6}$$

This can be done using iterated linear programming (LP). This algorithm was first proposed by Jowitt and Xu [6] and then refined by Pezzinga and Gueli [15] and by Creaco and Pezzinga [17,18] through the implementation of an underrelaxation method for setting update throughout the iterations.

The total daily leakage volume can be obtained as the sum of the leakage volumes $W_{L,j}$ at each time slot:

$$W_L = \sum_{i=1}^{n_{\Delta t}} W_{L,j}. \tag{7}$$

2.2. Optimal Location through Sequential Addition of Valves (SA)

The algorithm based on the sequential addition of valves (SA) was proposed by Pezzinga and Gueli [15]. In other words, SA is based on the optimal location of each single valve one after the other, independently from the potential benefit of the remaining ones. First, a maximum number N_{max} of valves installable in the network has to be fixed. The total number of steps in the algorithm is then N_{max} + 1. At Step 0 of this algorithm, the WDN has no control valves. At Step 1, the optimal location of 1 valve is searched for among the available locations in the WDN, i.e., all the n_p pipes. To this end, the control valve is placed inside the WDN model at one potential location at a time. The algorithm described in Section 2.1 is applied, enabling assessment of the installation cost of the system (or rather the total number of control valves as a surrogate, banally coinciding with the step of SA) and assessment of W_L. Then, the most beneficial valve, which yields the largest W_L reduction compared to the no valve scenario, is detected. At Step 2, the optimal location of two control valves is searched for, while keeping the first optimal control valve obtained from Step 1 installed in the WDN. The second valve for the optimal combination of two valves is searched for among the available locations, i.e., the $n_p - 1$ remaining pipes. The algorithm of Section 2.1 is applied considering $n_p - 1$ combinations of two control valves and the most beneficial one is detected in terms of W_L reduction compared to the 1 valve scenario. Other steps of SA can be carried out up to N_{max}, always considering the following basic assumption: at the generic Step N_{val}, the optimal combination of N_{val} valves is searched for, while containing the optimal combination of $N_{val} - 1$ valves detected at the previous step. Considering the number N_{max} of control valves installable in the WDN, SA would require the following number C_v^* of locations of control valves to be evaluated with the algorithm described in Section 2.1:

$$C_v^* = N_{max} n_p - \frac{N_{max}(N_{max} - 1)}{2}. \tag{8}$$

C_v^* is much smaller than the total enumeration C_v with no repetitions, which is given by:

$$C_v = \frac{N_p!}{N_{max}!(n_p - N_{max})!}. \tag{9}$$

Therefore, SA consists in a deterministically driven exploration of the research space of possible locations of control valves. The exploration of SA is expected to be very effective in the cases where the sequentially added valves do not affect each other. Otherwise, in those cases where the non-linearities of the problem are high, the effectiveness of SA is expected to diminish.

By plotting the W_L values obtained through SA as a function of N_{val}, with $0 \leq N_{val} \leq N_{max}$, a Pareto front of optimal solutions with $0 \leq N_{val} \leq N_{max}$ is obtained.

2.3. Optimal Location through Multi-Objective Genetic Algorithm (GA)

The multi-objective Non-Dominated Sorted Genetic Algorithm II (NSGA II) [22] was used to optimize the optimal location of control valves. In this GA, the adaptive mutation operator described by Carvalho and Araujo [23] was implemented. The decisional variables are encoded in individuals (i.e., solutions), with several genes equal to the number n_p of potential locations of control valves. Each gene can take on two possible values, 0 and 1, which indicate the absence and presence of the control valve at the generic pipe, respectively. The number N_{val} of valves for the generic solution is banally obtained by summing up the gene values.

In GA, the number of population individuals and the total number of generations must be fixed in light of the trade-off between computation time (which is related to the total number of objective function evaluations and is then a function of the available computation capabilities) and accuracy of the results. Incidentally, the total number of function evaluations is given by the total number of population individuals times the total number of generations.

The various individuals of the GA, which represent various locations of control valves installed in the network, are compared based on their fitness, composed of two objective functions evaluated through the algorithm described in Section 2.1, i.e., cost, or rather number of control valves, and W_L. Both the objective functions need to be minimized inside the GA. Similar to the algorithm for SA, the results of the multi-objective optimization are Pareto fronts. Unlike SA, in which the Pareto fronts feature a maximum number of valves equal to N_{max} depending on the number of SA steps, the GA Pareto fronts are made up of solutions with N_{val} ranging from 0 to n_p.

GA carries out a probabilistically driven exploration of the research space. Unlike SA, GA is theoretically able to explore all the possible locations.

3. Application

3.1. Case Studies

The application of this work concerned two case studies (Figure 1a,b, respectively). The first is the skeletonized WDN of Santa Maria di Licodia, a small town in Sicily, Italy [24]. The use of a skeletonized WDN is not expected to undermine the validity of the results since the pipes with larger diameter, which compose the WDN skeleton, are typically the best candidates for the installation of control valves.

The network layout is made up of $n_n = 34$ nodes (of which $n = 32$ with unknown head and $n_0 = 2$ source nodes with fixed head, i.e., Nodes 33 and 34) and of $n_p = 41$ pipes. The features of the WDN demanding nodes in terms of ground elevation z, mean daily demand d, and desired pressure head h_{des} for full demand satisfaction, are reported in Table 1. At the generic demanding node, the latter variable was set at the minimum value between 15 m and the daily lowest pressure head under uncontrolled conditions. This was done to avoid any pressure deficit increase at nodes that were pressure deficient ab initio. Table 2 reports the features of the pipes in terms of length L, diameter D and Strickler roughness coefficient k. The two source nodes, i.e., Nodes 33 and 34, have a ground elevation of 477.5 m a.s.l.

$N_{\Delta t} = 12$ 2-h-long time slots were used to represent the WDN daily operation. The patterns of the heads at the source nodes and of the hourly multiplying coefficient C_h for nodal demands are reported in Table 3.

As to leakage, two conditions were considered. In the former, values of the coefficient $C_L = 2.79 \times 10^{-8}$ m$^{0.82}$/s and of the exponent $n_{leak} = 1.18$ were assumed for all the pipes of the network. The values of C_L and n_{leak} lead to a daily leakage volume $W_L = 1242.6$ m^3, which is about 44% of the water volume leaving the source nodes, as evaluated in the real network. In the latter, the coefficient C_L was reduced to 0.85×10^{-8} m$^{0.82}$/s, obtaining a daily leakage volume $W_L = 397.7$ m^3 (about 20% of the water volume leaving the source nodes) without control valves.

The second case study is a district of the pipe network model used as benchmark in the Battles of Water Networks of the last WDSA conferences [19,20]. This district is made up of $n_n = 46$ nodes (of which $n = 45$ with unknown head and $n_0 = 1$ source node with assigned head, i.e., tank Node 46) and $n_p = 52$ pipes. The features of the district demanding nodes and pipes are provided by Creaco and Pezzinga [16], who also reported the values of the leakage exponent $n_{leak} = 0.9$ and of the leakage coefficient C_L ranging from 0.2 to 1 m$^{1.1}$/s in the various pipes. In this district, the leakage volume adds up to 118.7 m^3, which is 6.1% of the water volume entering the district, in the no valve scenario.

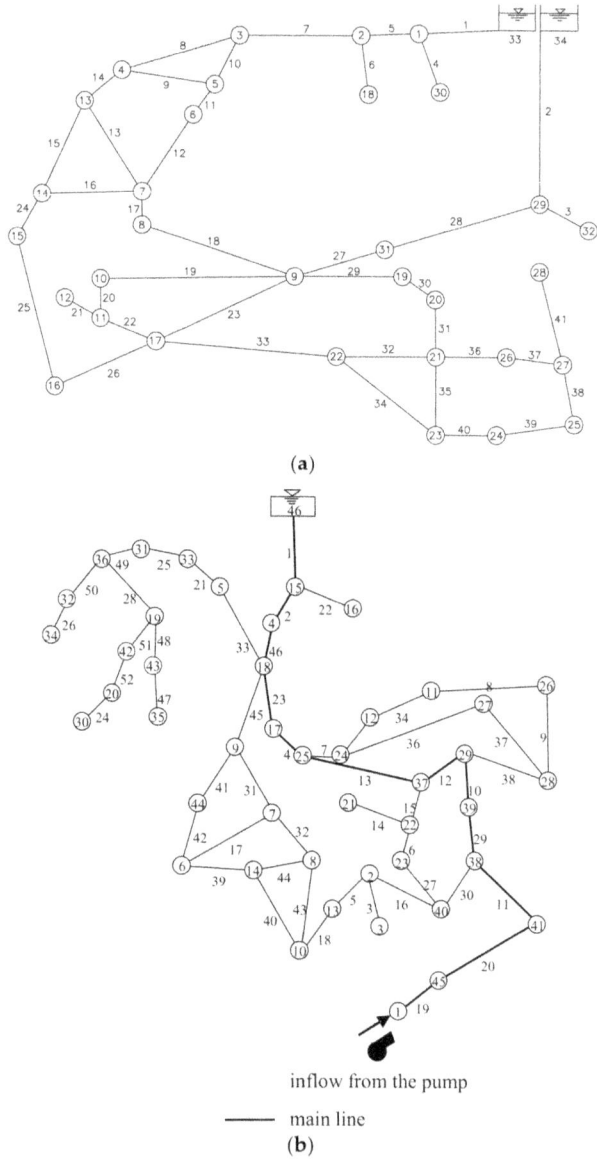

Figure 1. WDNs for the two case studies. Pipe numbers close to the pipes. Numbers of demanding nodes inside circles. (**a**) First case study has source Nodes 33 and 34. (**b**) Second case study has pump inflow at Node 1 and source Node 46. Main interconnection line is between the pump and source node.

Table 1. Features of network demanding nodes.

Node	z (m)	d (m^3/s)	h_{des} (m)	Node	z (m)	d (m^3/s)	h_{des} (m)
1	465	0	14.5	17	437.6	0.0007516	15
2	462	0	15	18	450	0.0006938	15
3	456	0.0001156	15	19	442	0.0011563	15
4	451.3	0.0001156	15	20	436.5	0.00075164	15
5	451	0.0002891	15	21	433.5	0.0008094	15
6	448.5	0.0002891	15	22	434	0.0008672	15
7	444	0.0006359	15	23	431.2	0.000925	15
8	446	0.0005781	15	24	436.8	0.0008672	15
9	445	0.0008672	15	25	435.8	0.0008672	15
10	442	0.0006938	15	26	438.6	0.0006938	15
11	438.6	0.0006359	15	27	440.3	0.0007516	15
12	437.5	0.0006938	15	28	430.1	0.0008672	15
13	448	0.0004047	15	29	465	0.0001734	11
14	435	0.0004625	15	30	445	0.0003469	15
15	441	0.0006938	15	31	454	0.0006938	15
16	394.8	0.0005781	15	32	435	0.0002313	15

Table 2. Features of network pipes in terms of length L, diameter D and Stricker roughness k.

Pipe	L (m)	D (mm)	k (m$^{1/3}$/s)	Pipe	L (m)	D (mm)	k (m$^{1/3}$/s)
1	352	250	75	22	110	125	65
2	314	175	65	23	214	150	65
3	1100	125	75	24	85	100	65
4	350	100	65	25	398	100	65
5	96	250	75	26	242	100	65
6	282	100	75	27	118	175	65
7	148	250	75	28	324	175	65
8	256	250	75	29	140	125	65
9	192	100	75	30	206	125	65
10	58	100	75	31	70	125	65
11	66	100	75	32	142	150	65
12	230	150	75	33	86	150	65
13	200	100	75	34	294	80	65
14	44	250	75	35	150	80	65
15	226	250	75	36	124	125	65
16	70	150	65	37	144	125	65
17	88	80	65	38	158	125	65
18	204	125	65	39	130	80	65
19	172	125	65	40	124	80	65
20	94	125	65	41	500	80	65
21	90	125	65				

Table 3. Temporal pattern of reservoir heads and of hourly demand coefficient C_h.

Time Slot (h)	Source 33–34 Head (m)	C_h (-)	Time Slot (h)	Source 33–34 Head (m)	C_h (-)
0–2	480.77	0.40	12–14	480.55	1.8
2–4	481.14	0.40	14–16	480.45	0.90
4–6	481.46	0.55	16–18	480.64	0.70
6–8	481.22	1.70	18–20	480.53	1.45
8–10	480.91	1.25	20–22	480.19	1.40
10–12	480.94	1.0	22–24	480.41	0.45

In the application, values equal to 105.8 m a.s.l. and 2.4 m, respectively, were assigned to the elevation and to the average water level of the source node. An inflow (that is a negative demand) takes place in correspondence to Node 1, at certain hours of the day, due to the activation of a pumping system that takes water from another district of the network. The detailed operation of this system, in terms of relationship between water discharge and head gain across the pump, is neglected in this work.

The district operation can be summarized in three time slots, for each of which the head of the source node, the inflow from the pump and the hourly demand coefficient for the demanding nodes were specified by Creaco and Pezzinga [18]. No valve installation is allowed in the main line (Figure 1b) that connects the pump with the tank node, to prevent any interferences with the filling/emptying process of the tank.

3.2. Results

In the application with SA [15] to the first case study, the optimal locations of up to $N_{max} = 10$ control valves were searched for. This required 365 objective function evaluations.

In the application with GA [17,18], a population of 50 individuals and a total number of 50 generations, corresponding to total number of 2500 objective function evaluations, were used. A healthy initial population was generated in GA to guarantee high computation efficiency at the end of the optimization.

The applications of this work were performed in the Matlab(R) 2011b environment by making use of a single processor of an Intel(R) Core(TM) i7-7700 3.60 GHz unit.

The tradeoff curves of W_L as a function of N_{val} obtained with SA and GA in the first leakage condition are reported in Figure 2. As for GA, two Pareto fronts are shown, the final one (after 50 generations) and that after seven generations. The latter is associated with 350 objective function evaluations (very close to SA). Though GA explored solutions with N_{val} values up to n_p (see Section 2.3), only the solutions with $N_{val} \leq 10$ were reported in the graph for a consistent comparison with SA. This graph shows that the curve of SA and the final curve of GA have identical trend up to $N_{val} = 3$. To the right of $N_{val} = 3$, the curve obtained with GA dominates that of SA, in that it provides lower values of W_L for each value of N_{val}. Furthermore, the distance between the two curves increases as N_{val} grew. This seems to suggest that, for low number of N_{val}, SA can provide identical or close results to those of GA. Conversely, for high number of N_{val}, the better performance of GA stands out, due to the wider exploration of the research space. However, this comes at an about seven times larger computational cost. As was expected, the curve of GA after seven generations features worse solutions than the final curve of GA to the right of $N_{val} = 3$. The curve of GA after seven generations enables the results of GA to be compared with those of SA, given the same computational budget. Interestingly, neither curve prevails in absolute terms. In fact, identical solutions are observed up to $N_{val} = 3$. GA after seven generations is slightly better for $N_{val} = 7, 9$ and 10, while not offering any solutions for $N_{val} = 8$. SA, instead, is better for $N_{val} = 5$ and 6. Summing up the results in Figure 2, the better performance of GA stands out only with a higher computational budget.

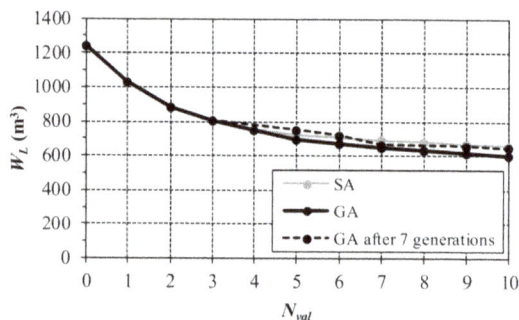

Figure 2. First case study. Daily leakage volume W_L as a function of N_{val} for the two algorithms.

An insight into the different results obtainable with SA and GA is provided in Table 4, which reports the optimal valve locations and the W_L values for the two algorithms. Whereas the valve positions suggested by SA and GA are the same up to N_{val} = 4, the two algorithms behave differently starting from the optimal location of five valves. In fact, at Step 5, SA suggests adding valve in Pipe 33 to the optimal location of N_{val} = 4 valves, in which the valve-fitted pipes are 3, 7, 14 and 27. Instead, for the sake of optimality, GA proposes valve elimination in Pipe 14 and valve insertion in Pipes 25 and 26, while moving from N_{val} = 4 to N_{val} = 5. The locations of SA and GA for N_{val} = 5 valves are shown in Figure 3. As Table 4 shows, this yields a 4.36% benefit of GA compared to SA, in terms of W_L. The percentage benefit of GA tends to grow up to almost 10% for N_{val} = 10.

Table 4. First case study. Optimal valve locations and daily leakage volumes W_L obtained with SA and GA. Benefits of GA in terms of W_L reduction.

N_{val}	Valve Locations with SA	W_L with SA (m³)	Valve Locations with GA	W_L with GA (m³)	Benefits of GA (%)
0	-	1243	-	1243	0.00
1	27	1029	27	1029	0.00
2	27,7	885	7,27	885	0.00
3	27,7,3	805	3,7,27	805	0.00
4	27,7,3,14	751	3,7,14,27	751	0.00
5	27,7,3,14,33	725	3,7,25,26,27	693	4.36
6	27,7,3,14,33,4	708	3,7,8,25,26,27	672	5.02
7	27,7,3,14,33,4,2	692	3,7,8,23,25,26,27	647	6.38
8	27,7,3,14,33,4,2,41	680	3,4,7,8,23,25,26,27	630	7.38
9	27,7,3,14,33,4,2,41,6	670	3,4,7,8,24,25,26,27,33	614	8.36
10	27,7,3,14,33,4,2,41,6,30	659	2,3,4,7,8,24,25,26,27,33	598	9.29

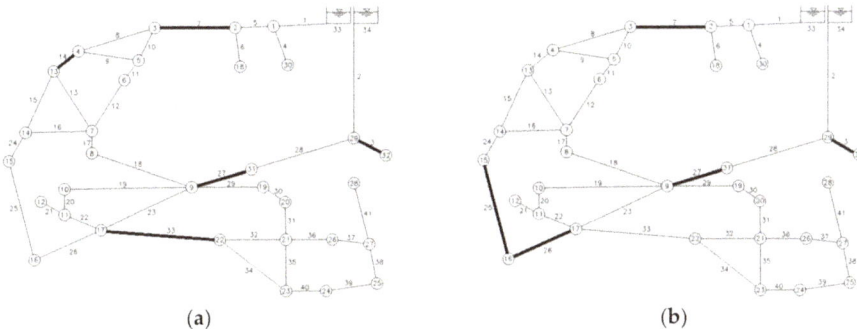

Figure 3. First case study. Optimal location of five valves for: (**a**) SA; and (**b**) GA. Valve locations indicated with thick lines.

Another example of the results obtained is provided in the graphs in Figure 4, associated with the optimal location of three valves, installed in Pipes 3, 7 and 27, respectively, for both SA and GA. Figure 4a shows the temporal pattern of V, optimized through the iterated LP at each time slot. Figure 4b shows the pressure head h_{down} at the downstream node, which can be used as the time varying setting to be prescribed if the water utility chooses to perform the pressure regulation by means of pressure reducing valves (PRVs). In fact, PRVs has the downstream pressure head as setting. This variable is always equal to h_{des} (that is 15 m) at the valve in Pipe 3 because the downstream node of Pipe 3, namely Node 32, is a terminal node of the network. It is always equal to h_{des} also for the valve in Pipe 7, though the downstream node of this pipe, namely Node 3, is not terminal. This happens because the descending topography downstream of Pipe 7 facilitates the meeting of the pressure requirements. The case of the valve in Pipe 27, which has Node 9 as downstream node, is similar. However, this valve cannot reduce h_{down} to h_{des} at nighttime, even if it is almost fully closed ($V \approx 0$).

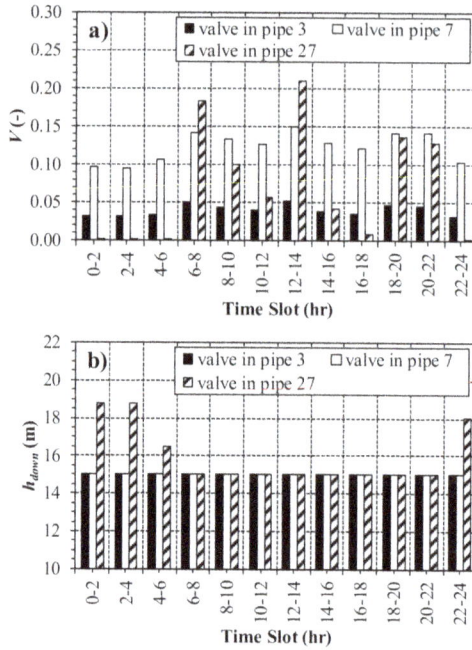

Figure 4. First case study. Optimal location of three valves. For each valve, trend of: V (**a**); and h_{down} (**b**) in the daily time slots.

An analysis should also be carried out concerning the possibility to convert the control valves proposed by the optimizer to isolation valves. This could be done when the optimal settings V are very close to 0 throughout the day. Furthermore, isolation valves could be closed in some pipes in parallel to the installed control valves if water flow is remarked to bypass the latter. However, neither situation was remarked to occur in the present case study. Furthermore, it must be noted that closing an isolation valve may decrease WDN redundancy and reliability.

The tradeoff curves of W_L as a function of N_{val} obtained with SA and GA in the second leakage condition are reported in Figure 5. This figure leads to similar remarks as Figure 2, as far as the comparison of the two algorithms is concerned.

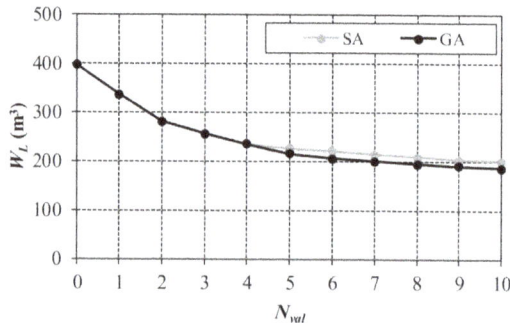

Figure 5. First case study under modified leakage conditions. Daily leakage volume W_L as a function of N_{val} for the two algorithms.

Table 5 gives some insight into optimal valve locations, W_L values and benefits of GA compared to SA. The comparison of Table 5 (low starting leakage) and Table 4 (high starting leakage rate) points out that the reduction in the starting leakage has the following minor effects:

- The optimal valve locations change for $N_{val} > 3$ (e.g., for $N_{val} = 4$, the optimal valve locations with GA are 3-7-24-27 and 3-7-14-27 in Table 5 and in Table 4, respectively).
- The benefits of GA in Table 4 tend to grow with N_{val} increasing, whereas the values in Table 5 tend to stabilize around 6.5%.

Table 5. First case study under modified leakage conditions. Optimal valve locations and daily leakage volumes W_L obtained with SA and GA. Benefits of GA in terms of W_L reduction.

N_{val}	Valve Locations with SA	W_L with SA (m³)	Valve Locations with GA	W_L with GA (m³)	Benefits of GA (%)
0	-	398	-	398	0.00
1	27	338	27	338	0.00
2	27-7	282	7-27	282	0.00
3	27-7-3	257	3-7-27	257	0.00
4	27-7-3-24	235	3-7-24-27	235	0.00
5	27-7-3-24-8	228	3-7-25-26-27	217	4.97
6	27-7-3-24-8-23	221	3-8-10-25-26-27	206	6.73
7	27-7-3-24-8-23-20	215	3-8-10-23-25-26-27	201	6.33
8	27-7-3-24-8-23-20-2	209	3-4-8-10-23-25-26-27	196	6.47
9	27-7-3-24-8-23-20-2-4	204	3-4-8-10-20-23-25-26-27	191	6.54
10	27-7-3-24-8-23-20-2-4-41	200	3-4-8-10-20-23-24-25-26-27	187	6.68

The applications to the second case study were carried out similarly to the first case study. The results are reported in Figure 6 and Table 6. The former reports the Pareto fronts obtained through SA and GA, whereas the latter reports optimal valve locations and W_L provided by the two algorithms, as well as the benefits of GA. Similar to the first case study, these benefits stand out only for $N_{val} \geq 5$. However, they are smaller (always below 4.32%), due to the different structure of the WDN. In fact, while the network of Santa Maria di Licodia is very interconnected, the network of the second case study is made up of quite independent branch structures fed by the main line, along which valves cannot be installed. This attenuates the non-linearities pertinent to optimal valve location.

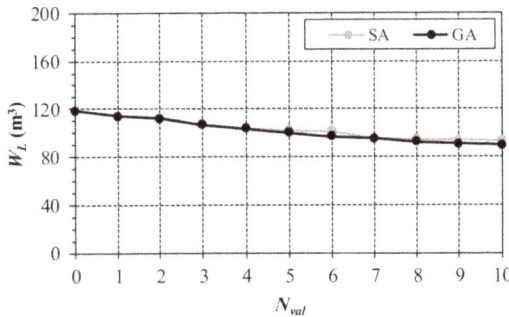

Figure 6. Second case study. Daily leakage volume W_L as a function of N_{val} for the two algorithms.

Table 6. Second case study. Optimal valve locations and daily leakage volumes W_L obtained with SA and GA. Benefits of GA in terms of W_L reduction.

N_{val}	Valve Locations with SA	W_L with SA (m³)	Valve Locations with GA	W_L with GA (m³)	Benefits of GA (%)
0	-	119	-	119	0.00
1	49	114	49	114	0.00
2	49-45	113	15-45	112	1.14
3	49-45-15	107	15-25-45	107	0.00
4	49-45-15-5	103	5 -15-45-49	103	0.00
5	49-45-15-5-27	102	7-15-38-45-49	100	1.85
6	49-45-15-5-27-7	101	5 -7-15-25-38-45	96	4.32
7	49-45-15-5-27-7-38	95	5-7-25-32-38-39-45	95	0.09
8	49-45-15-5-27-7-38-52	94	5-7-15-32-38-39-45-49	92	2.61
9	49-45-15-5-27-7-38-52-33	94	5-7-15-25-27-32-38-39-45	90	3.92
10	49-45-15-5-27-7-38-52-33-30	93	5-7-15-27-32-33-38-39-45-49	89	4.05

4. Conclusions

In this work, two algorithms for the optimal location of control valves, aimed at reducing service pressure and leakage in water distribution networks (WDNs), were compared. The former algorithm is deterministic and explores the research space of possible locations by adding one control valve at each step (sequential addition of valves SA). This exploration results in a significant reduction in the research space. The latter algorithm is a multi-objective genetic algorithm (GA), which explores, using a probability driven approach, the whole research space without restraints. The applications to skeletonized WDNs show that the two algorithms give identical results for a small number of installed control valves. However, GA outperforms SA when the number of installed valves increases. This may be because the non-linear effects of the optimization problem, totally neglected by SA, become more and more predominant as the number of installed valves grows. The benefits of GA are larger in the very interconnected WDNs, in which the non-linearities are emphasized. However, it must be noted that SA has a much smaller computational overhead and may then be preferable in the case of overly large networks. Furthermore, valve cost increase typically discourages water utilities from installing too many valves in the network. However, an advantage of GA lies in the possibility of accounting for other issues in the evaluation of the fitness of the generic solution. As an example, different and more complete objective functions could be easily and more rigorously considered inside GA, such as a relationship to express valve cost as a function of the diameter of the valve-fitted pipe. Furthermore, the closure of isolation valves, which are already available in WDNs to isolate segments [25], can be accounted for in GA to improve leakage reduction, as was shown in the works [17,18].

Acknowledgments: This research was conducted using the funds supplied by the University of Catania and by the University of Pavia.

Author Contributions: Both authors contributed equally to the paper.

Conflicts of Interest: The authors declare no conflict of interest.

References

1. Farley, M.; Trow, S. *Losses in Water Distribution Networks*; IWA: London, UK, 2003.
2. Puust, R.; Kapelan, Z.; Savic, D.A.; Koppel, T. A review of methods for leakage management in pipe networks. *Urban Water J.* **2010**, *7*, 25–45. [CrossRef]
3. Vicente, D.; Garrote, L.; Sánchez, R.; Santillán, D. Pressure Management in Water Distribution Systems: Current Status, Proposals, and Future Trends. *J. Water Resour. Plan. Manag.* **2016**, *142*, 04015061. [CrossRef]
4. Sinagra, M.; Sammartano, V.; Morreale, G.; Tucciarelli, T. A new device for pressure control and energy recovery in water distribution networks. *Water* **2017**, *9*, 309. [CrossRef]
5. Carravetta, A.; Del Giudice, G.; Fecarotta, O.; Ramos, H.M. Energy Production in Water Distribution Networks: A PAT Design Strategy. *Water Resour. Manag.* **2012**, *26*, 3947–3959. [CrossRef]
6. Jowitt, P.W.; Xu, C. Optimal Valve Control in Water-Distribution Networks. *J. Water Resour. Plan. Manag.* **1990**, *116*, 455–472. [CrossRef]

7. Campisano, A.; Creaco, E.; Modica, C. RTC of valves for leakage reduction in water supply networks. *J. Water Resour. Plan. Manag.* **2010**, *136*, 138–141. [CrossRef]
8. Vairavamoorthy, K.; Lumbers, J. Leakage reduction in water distribution systems: Optimal valve control. *J. Hydraul. Eng.* **1998**, *124*, 1146–1154. [CrossRef]
9. Reis, L.; Porto, R.; Chaudhry, F. Optimal location of control valves in pipe networks by genetic algorithm. *J. Water Resour. Plan. Manag.* **1997**, *123*, 317–326. [CrossRef]
10. Araujo, L.; Ramos, H.; Coelho, S. Pressure control for leakage minimisation in water distribution systems management. *Water Resour. Manag.* **2006**, *20*, 133–149. [CrossRef]
11. Liberatore, S.; Sechi, G.M. Location and Calibration of Valves in Water Distribution Networks Using a Scatter-Search Meta-heuristic Approach. *Water Resour. Manag.* **2009**, *23*, 1479–1495. [CrossRef]
12. Ali, M.E. Knowledge-Based Optimization Model for Control Valve Locations in Water Distribution Networks. *J. Water Resour. Plan. Manag.* **2015**, *141*, 04014048. [CrossRef]
13. Covelli, C.; Cozzolino, L.; Cimorrelli, L.; Della Morte, R.; Pianese, D. Optimal Location and Setting of PRVs in WDS for Leakage Minimization. *Water Resour. Manag.* **2016**, *30*, 1803–1817. [CrossRef]
14. De Paola, F.; Galdiero, E.; Giugni, M. Location and Setting of Valves in Water Distribution Networks Using a Harmony Search Approach. *J. Water Resour. Plan. Manag.* **2017**, *143*, 04017015. [CrossRef]
15. Pezzinga, G.; Gueli, R. Discussion of "Optimal Location of Control Valves in Pipe Networks by Genetic Algorithm.". *J. Water Resour. Plan. Manag.* **1999**, *125*, 65–67. [CrossRef]
16. Nicolini, M.; Zovatto, L. Optimal Location and Control of Pressure Reducing Valves in Water Networks. *J. Water Resour. Plan. Manag.* **2009**, *135*, 178–187. [CrossRef]
17. Creaco, E.; Pezzinga, G. Multi-objective optimization of pipe replacements and control valve installations for leakage attenuation in water distribution networks. *J. Water Resour. Plan. Manag.* **2015**, *141*, 04014059. [CrossRef]
18. Creaco, E.; Pezzinga, G. Embedding Linear Programming in Multi Objective Genetic Algorithms for Reducing the Size of the Search Space with Application to Leakage Minimization in Water Distribution Networks. *Environ. Model. Softw.* **2015**, *69*, 308–318. [CrossRef]
19. Creaco, E.; Alvisi, S.; Franchini, M. Multi-step approach for optimizing design and operation of the C-Town pipe network model. *J. Water Resour. Plan. Manag.* **2016**, *142*, C4015005. [CrossRef]
20. Giustolisi, O.; Berardi, L.; Laucelli, D.; Savic, D.; Walski, T. Battle of background leakage assessment for Water Networks (BBLAWN) at WDSA conference 2014. *Procedia Eng.* **2014**, *89*, 4–12. [CrossRef]
21. Creaco, E.; Franchini, M. Comparison of Newton-Raphson Global and Loop Algorithms for Water Distribution Network Resolution. *J. Hydraul. Eng.* **2014**, *140*, 313–321. [CrossRef]
22. Deb, K.; Pratap, A.; Agarwal, S.; Meyarivan, T. A fast and elitist multiobjective genetic algorithm NSGA-II. *IEEE Trans. Evolut. Comput.* **2002**, *6*, 182–197. [CrossRef]
23. Carvalho, A.G.; Araujo, A.F.R. Improving NSGA-II with an adaptive mutation operator. In Proceedings of the 11th Annual Conference Companion on Genetic and Evolutionary Computation Conference: Late Breaking Papers, Montreal, QC, Canada, 8–12 July 2009; ACM: New York, NY, USA, 2009; pp. 2697–2700.
24. Pezzinga, G. Procedure per la riduzione delle perdite mediante il controllo delle pressioni. In *Ricerca e Controllo Delle Perdite Nelle Reti di Condotte. Manuale Per una Moderna Gestione Degli Acquedotti*; Brunone, B., Ferrante, M., Meniconi, S., Eds.; CittàStudiEdizioni: Novara, Italy, 2008. (In Italian)
25. Creaco, E.; Franchini, M.; Alvisi, S. Evaluating water demand shortfalls in segment analysis. *Water Resour. Manag.* **2012**, *26*, 2301–2321. [CrossRef]

Article

Exploring Numerically the Benefits of Water Discharge Prediction for the Remote RTC of WDNs

Enrico Creaco [1,2,3]

[1] Dipartimento di Ingegneria Civile e Architettura, University of Pavia, Via Ferrata 3, 27100 Pavia, Italy; creaco@unipv.it

[2] School of Civil, Environmental and Mining Engineering, University of Adelaide, Adelaide SA 5005, Australia

[3] College of Engineering, Mathematics and Physical Sciences, University of Exeter, Exeter EX4, UK

Received: 20 October 2017; Accepted: 7 December 2017; Published: 9 December 2017

Abstract: This paper explores numerically the benefits of water discharge prediction in the real time control (RTC) of water distribution networks (WDNs). An algorithm aimed at controlling the settings of control valves and variable speed pumps, as a function of pressure head signals from remote nodes in the network, is used. Two variants of the algorithm are considered, based on the measured water discharge in the device at the current time and on the prediction of this variable at the new time, respectively. As a result of the prediction, carried out using a polynomial with coefficients determined through linear regression, the RTC algorithm attempts to correct the expected deviation of the controlled pressure head from the set point, rather than the currently measured deviation. The applications concerned the numerical simulation of RTC in a WDN, in which the nodal demands are reconstructed stochastically through the bottom-up approach. The results prove that RTC benefits from the implementation of the prediction, in terms of the closeness of the controlled variable to the set point and of total variations of the device setting. The benefits are more evident when the water discharge features contained random fluctuations and large hourly variations.

Keywords: valve; pump; real time control; pressure; water distribution modelling; leakage

1. Introduction

Nowadays, water utility managers often choose to perform active pressure control in their water distribution networks (WDNs) due to the many associated benefits [1], which include leakage and pipe burst reduction and the extension of network infrastructure useful life [2]. Another benefit of pressure control lies in the possibility of recovering energy from the WDN through turbines or similar devices [3,4]. After the WDN is subdivided in pressure zones [5], active pressure control requires a suitable pressure control device to be installed in the pipe(s) connecting each pressure zone to its source(s). Recent studies have proven that active pressure control can be more cost effective when the device setting is controlled in real time, in order to meet WDN demand variations in time [6]. In this context, the device controlled in real time can be a pressure control valve [7] or a variable speed pump [8], depending on whether the source node is under pressure surplus or deficit conditions compared to the pressure zone it feeds.

Remote real time control (RTC) is a kind of control in which the pressure head is monitored at remote critical node(s) in the WDN. By making use of suitable algorithms operating on the pressure head measurements, a programmable logic controller sets the new suitable device setting to maintain the minimum desired pressure at the remote node(s). In the scientific literature of WDNs, the first study on service pressure RTC was carried out by Campisano et al. [9], followed by many others [10–15]. Whereas the proportional algorithms of Campisano et al. [9,10,12] only used the pressure head measurement at the critical node, Creaco and Franchini [11] set up a more effective algorithm that also makes use of the water discharge measurement in the pipe fitted with the control device. Like the algorithm used by Campisano et al. [9,10,12], the algorithm of Creaco and Franchini [11] has a parameter that needs to be tuned to ensure the maximum regulation performance. Later, a parameter-less algorithm with only slightly worse performance than the algorithm of Creaco and Franchini [11] was proposed for the RTC of control valves [13] and variable speed pumps [14]. The algorithm of Page et al. [13,14] has the novel aspect of making use of the water discharge prediction in the pipe fitted with the control device. The prediction is carried out through water discharge measurements at small temporal steps (smaller than the RTC temporal step, which has order of magnitude of some minutes). However, a limit of this approach lies in the fact that it is effective only if the water discharge trend is smooth.

In this work, the water discharge prediction is implemented in a refined version of the algorithm of Creaco and Franchini [11]. Unlike the work of Page et al. [13,14], the prediction method operates on larger temporal steps than the control temporal step. Therefore, it is able to cope with the irregular trend of the pipe water discharge at fine temporal scale, due to the random nature of demand [15,16]. To show this, a stochastic bottom-up approach is used for nodal demand reconstruction in the applications, which concern the numerical simulation of RTC in a WDN.

2. Materials and Methods

In the following subsections, first the algorithm proposed by Creaco and Franchini [11] for control valve regulation is summarized, followed by its extension to variable speed pumps. Then, a novel upgrade of the algorithm is proposed, to include prediction at the new time. In both the cases of control valve and variable speed pump, the device setting is regulated in such a way to bring the pressure head at the control node close to the desired set point value. This node is selected as that which features the lowest pressure head among all the nodes sensitive to valve regulation [9].

2.1. Control Valve Regulation

The algorithm is described with reference to the schematic on the left side of Figure 1, in which a control valve is positioned in a pipe upstream from the control node B. At current time t_1 (s), let the water discharge in the pipe, temporally averaged over the control temporal step Δt (s), be equal to Q (m^3/s). The valve closure setting is then assumed to be equal to α_1 (-), which can range from 0 (fully open valve) to 1 (fully closed valve). If $\zeta(\alpha)$ is the valve curve, ζ_1 (-) is the local head loss coefficient associated with α_1. As shown in [5], the corresponding local head loss Δh_1 (m) is then equal to:

$$\Delta h_1 = \frac{\zeta_1 Q^2}{2g A^2},$$ (1)

where A (m^2) is the pipe inner cross section area and $g = 9.81$ m/s^2 is the gravity acceleration. In the downstream pipe end B, a temporally averaged pressure head value h_B (m), far from the set point value h_{sp} (m) by a quantity e (m), is observed. Incidentally, positive and negative values of e indicate larger and smaller values of h_B (m) than h_{sp}, respectively.

In order to bring the pressure head at node B close to the value h_{sp} at new time t_2 (s) (with $t_2 = t_1 + \Delta t$), the valve has to be regulated in such a way that the new head loss Δh_2 (m) is equal to:

$$\Delta h_2 = \Delta h_1 + e.$$ (2)

By making use of the relationship between head loss and head loss coefficient in Equation (1) and considering Equation (2), the head loss coefficient ζ_2 (-) at time t_2 has to be equal to:

$$\zeta_2 = \frac{2g A^2}{Q^2}(\Delta h_1 + e),$$ (3)

which can be rewritten as:

$$\xi_2 = \xi_1 + \frac{2g\,A^2}{Q^2}e. \tag{4}$$

In order to be able to tune the promptness of the control algorithm as well as to account for the fact that the system in Figure 1 is a simplified representation of a WDN, a correction coefficient K (-) is inserted in Equation (4), yielding the following Equation (5):

$$\xi_2 = \xi_1 + K\frac{2g\,A^2}{Q^2}e. \tag{5}$$

Indeed, coefficient K enables modulating the pressure head deviation e to correct. By using the relationship $\alpha(\xi)$ of the control valve, the valve setting α_2 corresponding to ξ_2 can be easily obtained. The setting variation from α_1 to α_2 has to be limited by the maximum correction allowed by the valve shutter velocity [9].

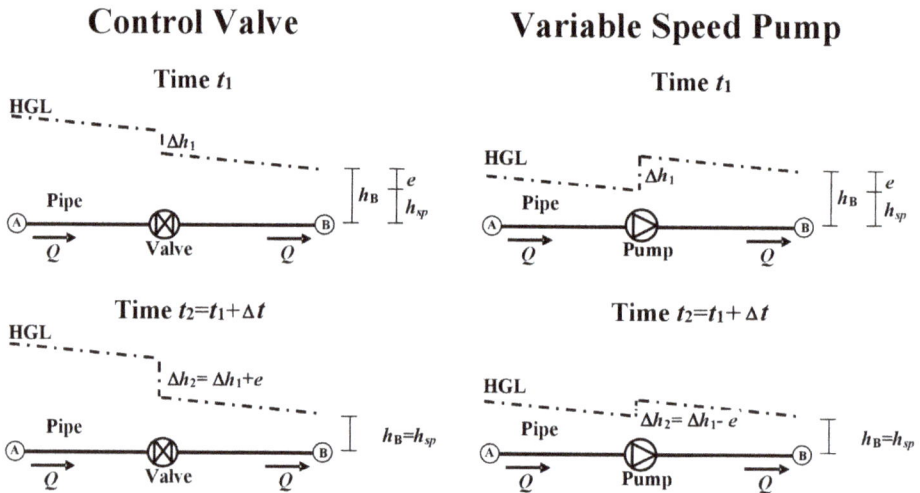

Figure 1. Schematic for the description of the operation of logic controller for the control valve (left) and for the variable speed pump (right): situation at the current (above) and new (below) times. HGL stands for head grade line.

2.2. Variable Speed Pump Regulation

The algorithm is described with reference to the schematic on the right side of Figure 1, in which a variable speed pump is positioned in a pipe upstream from the control node B. At current time t_1, let the water discharge in the pipe, temporally averaged over the temporal step Δt, be equal to Q. If β_1 (-), which can range from 0 (pump speed equal to 0) to 1 (maximum pump speed), is the current speed setting for the pump, the head Δh_1 provided by the pump is obtained through the pump curve:

$$\Delta h_1 = aQ^2 + bQ\beta_1 + c\beta_1^2, \tag{6}$$

where a (s^2/m^5), b (s/m^2) and c (m) are the pump curve coefficients. Equation (6) was obtained by considering that a second order curve can fit effectively the curve data of head and water discharge from pump catalogues, and by applying the affinity laws for pumps [5].

Let the temporally averaged pressure head h_B in the downstream pipe end B be far from the set point value h_{sp} by a quantity e.

In order to bring the piezometric height at node B close to the value h_{sp} at new time t_2, the pump has to be regulated in such a way that the head is equal to:

$$\Delta h_2 = \Delta h_1 - e. \tag{7}$$

An estimate for the new pump setting β_2 (-) can then be calculated starting from the following equation, derived from Equation (7) and from Equation (6):

$$c\beta_2^2 + bQ\beta_2 - bQ\beta_1 - c\beta_1^2 + e = 0. \tag{8}$$

The positive solution for β_2 in the previous equation is:

$$\beta_2 = \frac{-bQ + \sqrt{b^2 Q^2 - 4c\left(-b\beta_1 Q - c\beta_1^2 + e\right)}}{2c}. \tag{9}$$

A corrective coefficient K is introduced in this case as well, so that the formula for the estimation of the new pump setting is:

$$\beta_2 = \frac{-bQ + \sqrt{b^2 Q^2 - 4c\left(-b\beta_1 Q - c\beta_1^2 + Ke\right)}}{2c}. \tag{10}$$

The setting variation from β_1 to β_2 has to be limited by the maximum correction allowed by the variable speed drive.

2.3. Algorithm Refinement through State Prediction

As described in the sections above, variables Q and e appearing in Equations (5) and (10) are estimated at current time t_1. However, the algorithm can be improved if it is aimed at eliminating the expected pressure head deviation at new time t_2, rather than at eliminating the pressure head deviation at current time t_1. Therefore, reference should be made to the predicted water discharge $Q_{forecast}$ [m^3/s] and deviation $e_{forecast}$ [m]. $Q_{forecast}$ is the predicted water discharge at the new time; $e_{forecast}$, instead, is the deviation that would be expected at time t_2 if no setting variations were made from time t_1.

Unlike Page et al. [13,14], no water discharge measurements inside the control temporal step Δt are used in this work for water discharge prediction. This choice was made because the water discharge features, at small temporal scale, not negligible random fluctuations [15,16] that can distort the prediction. These fluctuations are due to the random nature of demand [15,16]. A further risk connected to the use of discharge measurements at small temporal scale, that is below the control temporal step Δt, is the presence of not temporally averaged measurement errors. Therefore, a prediction algorithm is set up to estimate $Q_{forecast}$, based on the value of the water discharge at time t_1 and at other previous times. At time t_1, based on the series of temporally averaged values $Q = Q(t_1)$, $Q(t_1 - \Delta t), Q(t_1 - 2\Delta t), \ldots , Q(t_1 - (N-1)\Delta t)$ derived from the available measurements, a regression is carried out to evaluate a smooth (fluctuation-free) trend of Q in the time interval from time $t_1 - (N-1)\Delta t$ to time t_1, with the following second order polynomial:

$$Q(t) = r_1 + r_2 t + r_3 t^2, \tag{11}$$

where coefficients r_1 (m^3/s) r_2 (m^3/s^2) and r_3 (m^3/s^3) are derived through the application of linear regression [17]. This polynomial is used for calculating $Q_{forecast}$, that is $Q(t_2 = t_1 + \Delta t)$.

At this stage, a caveat has to be made about the number N (-) of water discharge measurements to consider for estimating the coefficients of the polynomial in Equation (11). Indeed, this represents the number N of measurements that must be stored inside the logic controller to make the prediction. To estimate coefficients r_1, r_2 and r_3, a value of $N \geq 3$ must be considered. Incidentally, for $r = 3$, no best fit has to be searched for since only one parabola passes through three points. At each case study, an ad hoc analysis should be performed to understand the best window size N. As an explicative example of the prediction algorithm, let us consider a water discharge prediction at time t_1, in the context of RTC with $\Delta t = 180$ s (see graphs in Figure 2, in which dots and lines represent the measured values of Q and the polynomials obtained through regression, respectively). As graph a) highlights, if the water discharge trend is smooth (regular trend of the dots), as is the case with the aggregation of stochastically reconstructed demands from very numerous users, $N = 3$ already gives excellent prediction results. When the trend features visible random fluctuations (irregular trend of the dots), as is the case with the aggregation of stochastically reconstructed demands from few users (e.g., see Figure 2b), a larger value of N has to be considered to obtain a polynomial that reflects reliably the global temporal trend of Q, though failing to catch its random fluctuations. A small value of N (e.g., $N = 3$), instead, makes the water discharge prediction too dependent on random fluctuations. At each case study, the optimal window size must be assessed with the objective to maximize the overall performance of water discharge prediction and, therefore, of RTC. In this context, the closeness of the controlled variable to the set point can be considered as an indicator of the RTC performance for the choice of N, as is shown in Appendix A.

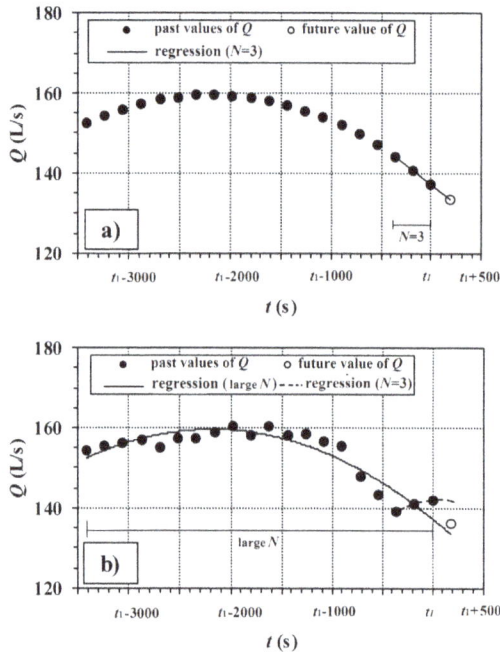

Figure 2. Example of application of the polynomial regression method for water discharge prediction in smooth (**a**) and irregular (**b**) demand trends.

As for $e_{forecast}$, which is associated with $Q_{forecast}$, the two following expressions can be used in the case of control valve or variable speed pump, respectively:

$$e_{forecast} = e - \zeta_1 \frac{\left(Q^2_{forecast} - Q^2\right)}{2g\,A^2}. \tag{12}$$

$$e_{forecast} = e + a\left(Q^2_{forecast} - Q^2\right) + b(Q_{forecast} - Q)\beta_1. \tag{13}$$

The formulas in Equations (12) and (13) reflect how the pressure head at the critical node is expected to vary if the water discharge in the device changes from the value Q to the value $Q_{forecast}$, in the absence of device setting variations. Potentially, two additional terms could be inserted in the right side of the expressions (12) and (13): the former associated with the predicted pressure-head variation upstream from the device (i.e., at the source) from time t_1 to time t_2; the latter associated with the predicted head loss variation in the WDN from time t_1 to time t_2. However, these terms are neglected in the present work because they are usually smaller than the contribution related to the variation in head loss (Equation (12)) or in head gain (Equation (13)) across the device. Furthermore, the assessment of these terms would require other variables to be monitored in the WDN. Ultimately, the presence of the corrective term K in Equation (5) and in Equation (10) enables taking account of all the effects not explicitly considered in the analysis.

3. Results

The applications concerned the numerical simulation of a network in Northern Italy, which serves about 30,000 inhabitants and has already been used for research in the field of pressure control [6,11,15]. The skeletonized layout is reported in Figure 3.

The network has a single source node, 26 demanding nodes with ground elevation of 0 m a.s.l. and 32 pipes. The uniform ground elevation and the size make the network eligible for being considered as a single pressure zone.

More details about characteristics of the pipes, in terms of pipes and length, can be found in the referenced work.

Two scenarios were considered hereinafter. In scenario 1, the source node is considered to have a constant head of 40 m a.s.l. Furthermore, a DN350 plunger control valve is inserted in pipe 26-11, which connects the source node to the rest of the network. The valve curve $\zeta(\alpha)$ considered for the plunger valve has the following form [10]:

$$\zeta = 10^{c_1 - c_2 \log_{10}(1-\alpha)}, \tag{14}$$

where $c_1 = 0.75$ and $c_2 = 3.25$ are the coefficients calculated to best fit the data provided by a valve manufacturer. In Equation (14), α was allowed to range from 0 (fully open) to 0.95 (almost fully closed).

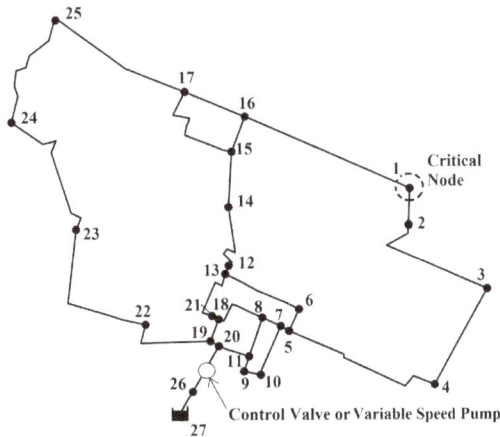

Figure 3. Water distribution network considered for the applications.

In scenario 2, the source node is considered to have a constant head of 1 m a.s.l. and a variable speed pump is inserted in pipe 26-11. The curve used for this device is the following:

$$\Delta h = -250\,Q^2 - 20\,Q\beta + 48\beta^2, \tag{15}$$

with β allowed to range from 0.5 (minimum speed) to 1 (maximum speed).

In both scenarios, the critical node of the network, which features the lowest pressure head values along the day, is node 1. Therefore, node 1 was selected as control node. The RTC of the control valve and of the variable speed pump, in scenarios 1 and 2, respectively, was carried out to bring the pressure head at the control node to set point value $h_{sp} = 25$ m.

For each node, the demand trend was obtained through the bottom-up approach by applying the methodology described in [15,16], which is based on demand pulse generation through the Poisson model. In the present work, the same pulse features as those associated with Dutch households with two to three inhabitants (reported in [16]), in terms of duration, intensity, and duration/intensity correlation, were considered. After being generated for five consecutive days and for all the network nodes, the demand pulses were aggregated over $\Delta t = 180$ s long temporal steps, to correspond to the temporal step adopted for RTC and for the extended period simulation [18] of the network. At this stage, a remark must be made about the choice of $\Delta t = 180$ s as control temporal step for RTC. In fact, this choice was made based on the findings in [15], which proved that 180 s guarantees the best trade-off between closeness of the controlled variable to the set point and number of actuator setting variations in this case study.

Each instant of network operation was simulated through the global gradient algorithm [19]. Incidentally, the combination of extended period simulation with nodal demands reconstructed with the bottom-up approach at medium temporal steps represents the best combination to simulate the behavior of RTC [15] and WDNs [16] in real time, in the trade-off between accuracy of the results and computational efficiency.

At each node and at each instant of simulation, the outflow to the users was evaluated with the pressure-driven approach through the Wagner et al. [20] formula, taking h_{sp} as the pressure head for full demand satisfaction at all nodes. Furthermore, leakage q_{leak} (m^3/s) was related to the nodal pressure head h through the Tucciarelli et al. [21] formula:

$$q_{\text{leak}} = \alpha_{\text{leak}} h^\gamma \frac{\sum L}{2}, \tag{16}$$

where $\sum L$ (m) is the total length of the pipes connected to the node. Furthermore, α_{leak} (m$^{2-\gamma}$/s) and γ (-) are the leakage coefficient and exponent, respectively. While γ depends on pipe material and leak opening shape [22], α_{leak} depends on the number of leak openings along the pipe and then grows with pipe age. In this work, these parameters were assumed uniform over the network. γ was set to 1 (typical value for plastic pipes) while α_{leak} was set to 9.4×10^{-9} m/s in order to have in the first scenario a leakage percentage rate close to 20% of the total outflow from the source, in the case of fully open control valve.

The total network demand D (L/s) trend at the $\Delta t = 180$ s temporal scale obtained with the bottom-up reconstruction is reported in graph a) in Figures 4 and 5 for the five consecutive days of network operation. These graphs show that, despite the spatial and temporal aggregation, demand random fluctuations are still present, especially under low demand conditions. Furthermore, each daily trend of demand is slightly different from the others. These aspects are due to the stochastic method used for demand generation, which obtains nodal demands as the superimposition of demand pulses [15,16], consistently with real WDNs.

For analyzing the performance of RTC, two indices were used. The former, indicated as mean $|e|$, concerns the closeness of the controlled variable at the generic i-th time, that is the pressure head h_i at the control node, to the set point value h_{sp}. It is calculated as:

$$\text{mean}|e| = \frac{\sum_{i=1}^{N_{tot}} |h_i - h_{sp}|}{N_{tot}}, \tag{17}$$

where $N_{tot} = 2400$ is the number of temporal steps Δt in the five days of simulation.

The latter index, indicated as $\sum |\Delta \alpha|$ and $\sum |\Delta \beta|$ for the control valve and for the variable speed pump, respectively, is representative of the total variations of the device setting. In the two scenarios, it is calculated as:

$$\sum |\Delta \alpha| = \sum_{i=2}^{N_{tot}} |\alpha_i - \alpha_{i-1}|, \text{ and} \tag{18}$$

$$\sum |\Delta \beta| = \sum_{i=2}^{N_{tot}} |\beta_i - \beta_{i-1}|, \tag{19}$$

where α_i and β_i are the values of valve setting α and pump setting β, respectively, at the generic i-th time. Generally speaking, a good controller is expected to keep the controlled variable close to the set point with small device setting variations, to avoid the wear of the control device. Therefore, it will feature a low value of both indices.

For both control valves (Equation (5)) and pumps (Equation (10)), the coefficient K of the logic controller was calibrated in such a way as to minimize mean $|e|$.

Globally, 4 simulations were carried out: simulations 1a and 1b, associated with scenario 1, and simulations 2a and 2b, associated with scenario 2. Simulations 1a and 2a were carried out considering the original RTC algorithm of Creaco and Franchini [11] with no water discharge prediction, hereinafter indicated as LCa (logic controller a). In simulations 1b and 2b, the refined RTC algorithm fitted with the water discharge prediction, hereinafter indicated as LCb (logic controller b), was used. In the application, the polynomial regression useful for the demand prediction was carried out on $N = 20$ time instants, following a preliminary analysis the results of which are reported in Appendix A.

The setting variation speed of both the control valve and the variable speed pump was set in such a way as to allow the maximum setting variation (0–0.95 and 0.5–1 for the former and the latter device, respectively) inside the control temporal step $\Delta t = 180$ s.

The results of the simulations in terms of optimal K, mean $|e|$ and $\sum |\Delta \alpha|$ are reported in Tables 1 and 2 for the control valve and the variable speed pump, respectively.

The results in Table 1 highlight, for the control valve controlled in real time, the superiority of LCb, in terms of both mean $|e|$ and $\sum |\Delta \alpha|$. In fact, accounting for the water discharge prediction enabled the controlled variable to stay, on average, closer to the set point with smaller control valve setting variations.

The results in Table 2 highlights, also for the variable speed pump controlled in real time, the better performance of LCb.

Table 1. Results of simulations 1a and 1b related to scenario 1 (control valve), in terms of mean $|e|$ and $\sum |\Delta \alpha|$ for the optimal value of K. LCa (logic controller a); LCb (logic controller b).

Simulation	Logic Controller	K (-)	Mean $|e|$ (m)	$\sum |\Delta \alpha|$ (-)
1a	LCa	0.5	0.79	26.59
1b	LCb	0.5	0.70	25.11

Other results of simulations 1b and 2b (with LCb) are reported in Figures 4 and 5, respectively. In either Figure, graphs b and c report the trend of the device setting and of the controlled pressure head, respectively. Graph b) in Figure 4 shows, as expected, that the control valve tends to open (lower values of α, down to about 0.2) in correspondence to the morning, midday and evening demand peaks. Outside the peaks, instead, the control valve tends to close (higher values of α, up to about 0.9), in order to cause larger local head losses. Only minor oscillations are observed in the valve setting trend. Graph c shows that h is, on average, always close to the set point. However, up to 2.5 m large overshooting and undershooting are remarked. These oscillations take place due to the imperfect effectiveness of the prediction method. In fact, as was remarked above with reference to Figure 2b, the prediction method is not able to catch the random fluctuations of the water discharge, though being able to detect its overall trend.

Table 2. Results of simulations 2a and 2b related to scenario 2 (variable speed pump), in terms of mean $|e|$ and $\sum|\Delta\beta|$ for the optimal value of K. LCa (logic controller a); LCb (logic controller b).

Simulations	Logic Controller	K (-)	Mean $|e|$ (m)	$\sum|\Delta\beta|$ (-)
2a	LCa	1.4	0.25	10.66
2b	LCb	1.4	0.22	9.64

In this context, it has to be stressed that it is unfeasible to totally eliminate the random oscillations of the pressure head at a remote node. In fact, this would require the prediction/time step to be reduced. However, the control time step cannot be reduced below a threshold value, which accounts for signal propagation from the control node to the valve, as was stated in [15].

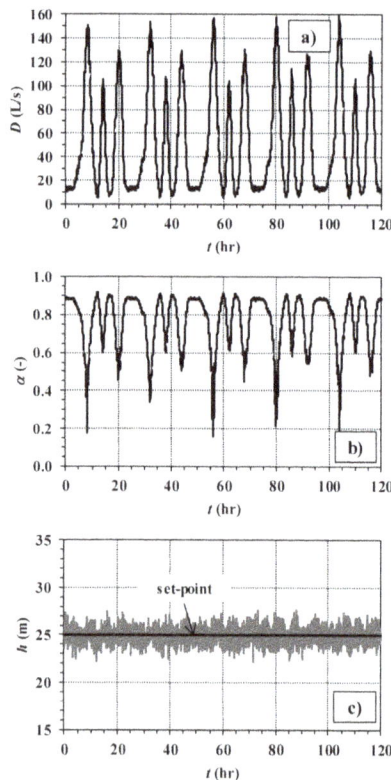

Figure 4. Real time control (RTC) of control valve through LCb. In the five days of network operation, trends of total network demand (**a**), device setting (**b**) and controlled pressure head (**c**).

The graphs in Figure 5 show that, contrary to α, the pump speed ratio β tends to increase during demand peaks and to decrease under low demand conditions. Furthermore, in the case of the variable speed pump, the oscillations of the controlled variable are smaller, especially under low demand conditions. This can be explained as follows. If we consider, for an assigned value of water discharge Q, the device settings bringing the controlled pressure head to the set point, a variation in Q, like that associated with a demand fluctuation, tends to produce a smaller change in terms of Δh for the pump (Equation (6)) than for the valve (Equation (1)). In other words, the pump tends to dampen the effects of water discharge variations better than the valve, especially under low-demand conditions (which require high valve closures and low pump speeds).

A sensitivity analysis is then carried out to investigate if the superiority of LCb persists when other values are used for the parameters in the bottom-up generation of nodal pulsed demands. In new simulations performed with the control valve, the pulse parameters associated with the Milford households [23] were used (see [16] for more details on the values of these parameters). Furthermore, a flatter trend was considered for the demand pattern throughout the day, leading to the total demand D shown in Figure 6a, where demand fluctuations are more evident than in Figures 4a and 5a. The optimization of parameter K yielded, for both LCa and LCb, the optimal value of 0.3, in correspondence to which the two algorithms produce an almost identical value of mean $|e| = 0.8$. However, LCb is still superior in terms of $\sum |\Delta \alpha|$ (9.42 for LCb vs. 11.09 for LCa). Graphs b and c in Figure 6 show the trend of the valve closure setting and of the controlled pressure head for LCb under conditions of modified demand.

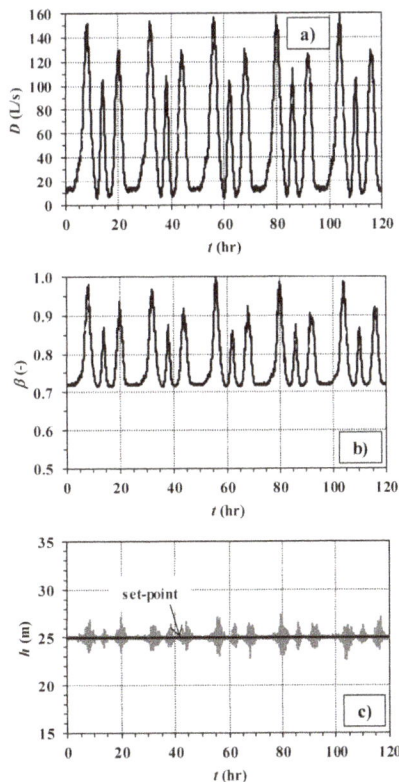

Figure 5. RTC of variable speed pump through LCb. In the five days of network operation, trends of total network demand (**a**), device setting (**b**) and controlled pressure head (**c**).

Overall, the advantages of LCb are less evident in the case of the modified pulse parameter values, due to the more evident presence of random fluctuations, uncaptured by the water discharge prediction method (see Figure 2), in the total network demand (compare Figure 6a with Figures 4a and 5a). This entails that, ceteribus paribus, the benefits of LCb will be more evident when the random fluctuations of demand are small compared to its hourly variations. Otherwise, the benefits of LCb will decrease due to the worse effectiveness of the prediction.

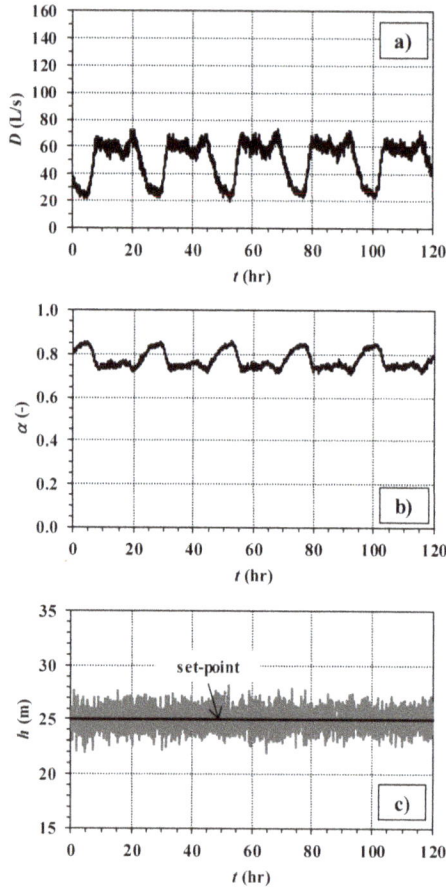

Figure 6. RTC of control valve through LCb under conditions of modified demand. In the five days of network operation, trends of total network demand (**a**), device setting (**b**) and controlled pressure head (**c**).

4. Conclusions

In this paper, two variants of the logic controller of [11] were numerically applied to the RTC of control valves and variable speed pumps, namely, its original version and the novel version fitted with the prediction of the water discharge through the device. Thanks to this prediction, the algorithm attempts to correct the expected deviation of the remotely controlled pressure head from the set point at the new time, rather than the measured one at the current time. In this work, water discharge prediction was carried out using a polynomial function constructed starting from the measured water discharges in the previous time instants. The results showed that this simple prediction method can represent the global temporal trend of the water discharge, even when nodal demands feature random fluctuations due to the stochastic reconstruction through the bottom-up approach. As a result, the performance of the RTC algorithm benefits from the implementation of the water discharge prediction method in terms of closeness of the controlled variable to the set point and of total variations of the device setting. The benefits stand out above all when the random fluctuations of demand are small compared to its hourly variations.

Future work will be dedicated to the setup of more complicated prediction methods, such as those based on probabilistic concepts. By enabling proper interpretation and representation of the random water discharge fluctuations, which arise because of the random nature of demand, these methods are expected to improve the effectiveness of the prediction and, as a result, the performance of RTC. Future work will also be dedicated to testing the RTC algorithms in other case studies, to explore the influence of network size and topography on the algorithm effectiveness.

Furthermore, in another development of this work, a different kind of device regulation will be used, which will consider an acceptable range of pressure heads at the critical node, rather than a single set point. This is expected to reduce the number of actuator setting variations and, ultimately, to extend the useful life of the device.

Acknowledgments: This research was conducted using the funds supplied by the University of Pavia.

Conflicts of Interest: The author declares no conflict of interest.

Appendix A

A preliminary analysis was carried out to select the optimal window size N on which to apply the regression in Equation (11). In this analysis, algorithm LCb was applied to the five days of network operation considering values of N ranging from 3 to 30. For each simulation in the analysis, mean $|e|$ was calculated. The graph in Figure A1 reports the trend of mean $|e|$ as a function of N (window size). The curve has a minimum at $N = 20$, indicating that, in the present case study, this is the optimal size of the window used for the prediction. This value was then adopted for all the simulations with LCb in the paper. However, as highlighted in the section entitled "Materials and Methods", the optimal value of N may be different in different case studies from that considered in this paper. Furthermore, it is expected to vary as a function of the control temporal step Δt. Therefore, an ad hoc analysis should be performed for assessing this parameter in each case study.

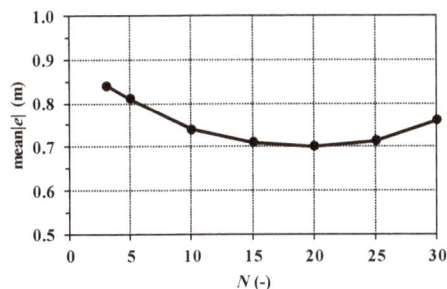

Figure A1. Trend of mean $|e|$ as a function of the window size N used for water discharge prediction.

References

1. Farley, M.; Trow, S. *Losses in Water Distribution Networks*; IWA: London, UK, 2003.
2. Vicente, D.; Garrote, L.; Sánchez, R.; Santillán, D. Pressure Management in Water Distribution Systems: Current Status, Proposals, and Future Trends. *J. Water Resour. Plan. Manag.* **2016**, *142*, 04015061. [CrossRef]
3. Fecarotta, O.; Aricò, C.; Carravetta, A.; Martino, R.; Ramos, H.M. Hydropower Potential in Water Distribution Networks: Pressure Control by PATs. *Water Resour. Manag.* **2014**, *29*, 699–714. [CrossRef]

4. Sinagra, M.; Sammartano, V.; Morreale, G.; Tucciarelli, T. A new device for pressure control and energy recovery in water distribution networks. *Water* **2017**, *9*, 309. [CrossRef]

5. Walski, M.; Chase, D.; Savic, D.; Grayman, W.; Beckwith, S.; Koelle, E. *Advanced Water Distribution Modelling and Management*; Haestad: Waterbury, CT, USA, 2003.

6. Creaco, E.; Walski, T. Economic Analysis of Pressure Control for Leakage and Pipe Burst Reduction. *J. Water Resour. Plan. Manag.* **2017**, *143*, 04017074. [CrossRef]

7. Fontana, N.; Giugni, M.; Glielmo, L.; Marini, G. Real Time Control of a PRV in Water Distribution Networks for Pressure Regulation: Theoretical Framework and Laboratory Experiments. *J. Water Resour. Plan. Manag.* **2018**, *144*, 04017075. [CrossRef]

8. Walski, T.; Creaco, E. Selection of Pumping Configuration for Closed Water Distribution Systems. *J. Water Resour. Plan. Manag.* **2016**, *142*, 04016009. [CrossRef]

9. Campisano, A.; Creaco, E.; Modica, C. RTC of valves for leakage reduction in water supply networks. *J. Water Resour. Plan. Manag.* **2010**, *136*, 138–141. [CrossRef]

10. Campisano, A.; Modica, C.; Vetrano, L. Calibration of proportional controllers for the RTC of pressures to reduce leakage in water distribution networks. *J. Water Resour. Plan. Manag.* **2012**, *138*, 377–384. [CrossRef]

11. Creaco, E.; Franchini, M. A new algorithm for the real time pressure control in water distribution networks. *Water Sci. Technol. Water Supply* **2013**, *13*, 875–882. [CrossRef]

12. Campisano, A.; Modica, C.; Reitano, S.; Ugarelli, R.; Bagherian, S. Field-Oriented Methodology for Real-Time Pressure Control to Reduce Leakage in Water Distribution Networks. *J. Water Resour. Plan. Manag.* **2016**, *142*, 04016057. [CrossRef]

13. Page, P.R.; Abu-Mahfouz, A.M.; Yoyo, S. Parameter-Less Remote Real-Time Control for the Adjustment of Pressure in Water Distribution Systems. *J. Water Resour. Plan. Manag.* **2017**, *143*, 04017050. [CrossRef]

14. Page, P.R.; Abu-Mahfouz, A.M.; Mothetha, M.L. Pressure Management of Water Distribution Systems via the Remote Real-Time Control of Variable Speed Pumps. *J. Water Resour. Plan. Manag.* **2017**, *143*, 04017045. [CrossRef]

15. Creaco, E.; Campisano, A.; Franchini, M.; Modica, C. Unsteady Flow Modeling of Pressure Real-Time Control in Water Distribution Networks. *J. Water Resour. Plan. Manag.* **2017**, *143*, 04017056. [CrossRef]

16. Creaco, E.; Pezzinga, G.; Savic, D. On the choice of the demand and hydraulic modeling approach to WDN real-time simulation. *Water Resour. Res.* **2017**, *53*. [CrossRef]

17. Freedman, D.A. *Statistical Models: Theory and Practice*; Cambridge University Press: Cambridge, UK, 2009.

18. Todini, E. A more realistic approach to the 'extended period simulation' of water distribution networks. In Proceedings of the Computing and Control for the Water Industry 2003, London, UK, 15–17 September 2003. [CrossRef]

19. Todini, E.; Pilati, S. A gradient algorithm for the analysis of pipe networks. In *Computer Application in Water Supply, System Analysis and Simulation*; Coulbeck, B., Choun-Hou, O., Eds.; John Wiley & Sons: London, UK, 1988; Volume 1, pp. 1–20.

20. Wagner, J.M.; Shamir, U.; Marks, D.H. Water distribution reliability: Simulation methods. *J. Water Resour. Plan. Manag.* **1988**, *114*, 276–294. [CrossRef]

21. Tucciarelli, T.; Criminisi, A.; Termini, D. Leak Analysis in Pipeline System by Means of Optimal Value Regulation. *J. Hydraul. Eng.* **1999**, *125*, 277–285. [CrossRef]

22. Van Zyl, J.E.; Cassa, A.M. Modeling elastically deforming leaks in water distribution pipes. *J. Hydraul. Eng.* **2013**, *140*, 182–189. [CrossRef]

23. Buchberger, S.G.; Carter, J.T.; Lee, Y.H.; Schade, T.G. *Random Demands, Travel Times and Water Quality in Dead-Ends*; AWWARF Rep. No. 294; American Water Works Association Research Foundation: Denver, CO, USA, 2003.

Article

PATs Operating in Water Networks under Unsteady Flow Conditions: Control Valve Manoeuvre and Overspeed Effect

Modesto Pérez-Sánchez [1], P. Amparo López-Jiménez [1] and Helena M. Ramos [2,*]

[1] Hydraulic and Environmental Engineering Department, Universitat Politècnica de València, 46022 Valencia, Spain; mopesan1@upv.es (M.P.-S.); palopez@upv.es (P.A.L.-J.)

[2] Civil Engineering, Architecture and Georesources Departament, CERIS, Instituto Superior Técnico, Universidade de Lisboa, 1049-001 Lisboa, Portugal

* Correspondence: hramos.ist@gmail.com; Tel.: +351-913582452

Received: 8 March 2018; Accepted: 9 April 2018; Published: 23 April 2018

Abstract: The knowledge of transient conditions in water pressurized networks equipped with pump as turbines (PATs) is of the utmost importance and necessary for the design and correct implementation of these new renewable solutions. This research characterizes the water hammer phenomenon in the design of PAT systems, emphasizing the transient events that can occur during a normal operation. This is based on project concerns towards a stable and efficient operation associated with the normal dynamic behaviour of flow control valve closure or by the induced overspeed effect. Basic concepts of mathematical modelling, characterization of control valve behaviour, damping effects in the wave propagation and runaway conditions of PATs are currently related to an inadequate design. The precise evaluation of basic operating rules depends upon the system and component type, as well as the required safety level during each operation.

Keywords: energy recovery systems; runaway conditions; unsteady flow; water hammer

1. Introduction

The need to increase the efficiency in pressurized water networks has allowed the development of new water management strategies in the last decades [1,2]. These strategies have focused on two different directions according to the water pressurized network type (i.e., pumped or gravity systems). In pump solutions, the efficiency increase in the network is directly correlated with the reduction of the manometric head [3,4], the correction of operation rules and the design of facilities (e.g., pump efficiency, leakage control and the establishment of optimum schedules) [5]. In gravity systems, the efficiency improvement is related to the reduction of the leakage level through the installation of pressure reduction valves [6–10]. Ramos and Borga (1999) proposed the replacement of pressure reduction valves (PRVs) by hydraulic machines, which could also generate energy [11]. These systems provide two benefits: on the one hand, PATs reduce the pressure in the system and therefore, the leakages are also reduced by the operation of PRVs; on the other hand, the generated energy contributes to the improvement of the energy balance of these water systems, increasing the efficiency in the water networks, as well as improving performance indicators [12]. PATs can be used in any pipe system with excess of flow energy being more suitable for: (1) water supply networks, (2) irrigation systems, (3) industry processes, (4) drainage or storm systems and (5) treatment plants or at the entrance of reservoirs/tanks. The range operation (i.e., flow and head) is high and depends on the selected machine (i.e., radial, axial, mixed and multistage). Commonly, the flow rate is between 1 and 100 l/s and the head rate oscillates between 1 and 80 m w.c. (meters water column) However, it is possible to reach higher flows and heads if the machines are installed in parallel or in series.

A deep analysis of the use of PATs in pipe systems, as well as the operation rate was described by [12]. Therefore, the success of PATs is related to the high operation rate and their combination (parallel or series), enabling the installation of these recovery machines where traditional turbines are not suitable.

Commonly, when replacing PRVs, a proposed hydraulic machine is a PAT [6]. Numerous researchers analysed the behaviour of these machines under steady flow conditions. A review of available technologies was conducted by different researchers [12–15]. A PAT analysis of performance and modelling was done on different hydraulic machines [14,16–19], while the computational analysis of these machines in a water distribution network was also studied [20–23]. The design of innovative strategies to maximize the recovered energy when the flows vary along a day, as well as their economic feasibility [24–27], in which the computed payback period achieved values between 2–12 years, depending on the system characteristics, was presented. Therefore, although the PAT installation is generally feasible, there are some cases in which the investment is economically unfeasible [28]. These strategies were applied to determine and maximize the theoretical recovered energy in both drinking and irrigation water systems [29,30]. The last case studies consider the significance of the flow change over time to predict the generated power in these facilities when they are installed in water systems [31]. The variability of the PATs performance as a function of flow, the maximization of the recovered energy that was developed using optimization procedures, and the economic analysis were successfully introduced in the analysis of a water pressurized system [32].

However, the study of the unsteady flow is poorly analysed in these systems and the installation of PATs encourages the need to know more about this subject. The transient analysis allows the estimation of the overpressures that could risk hydraulic facilities [33,34]. As a novelty, this research analyses the effective percentage of closure (effective %) in valve manoeuvres, the start-up and shutdown of radial and axial PATs with low inertia (i.e., of small sizes), as well as the runaway conditions induced by the overspeed effect through experimental data collection.

2. Material and Methods

2.1. Basic Hydraulic Modelling of the Transient Conditions

The unsteady flow can be analysed by a one-dimensional (1D) model type in pressurized pipe systems with higher length than diameter, using the mass and momentum conservation equations which are derived from the Reynolds transport theorem [35]. These principles are defined by differential Equations (1) and (2) [36–39]:

$$\frac{\partial H}{\partial t} + \frac{c^2}{gA}\frac{\partial Q}{\partial x} = 0 \tag{1}$$

$$\frac{\partial H}{\partial x} + \frac{1}{gA}\frac{\partial Q}{\partial t} + \frac{4\tau_w}{\rho g D} = 0 \tag{2}$$

where H is the piezometric head in m; t is the time in s; c is the pressure wave speed in m/s, which is defined by the Equation (3); g is the gravity acceleration in m/s^2; A is the inner area of the pipe in m^2; Q is the flow in m^3/s; x is the coordinate along the pipeline axis; τ_w is the shear stress at the pipe wall in N/m^2; ρ is the density of the fluid in kg/m^3; and D is the inner diameter of the pipe in m.

$$c = \sqrt{\frac{K}{\rho(1+(K/E)ps)}} \tag{3}$$

where K is the fluid bulk modulus of elasticity in N/m^2; E is the Young's modulus of elasticity of the pipe in N/m^2; and ps is the dimensionless parameter that takes into account the cross-section parameter of the pipe and supports constraint.

The considered assumptions applied in the classical, one dimensional, water hammer models are [37–39]:

- The flow is homogenous and compressible;

- The changes of density and temperature in the fluid are considered negligible when these are compared to pressure and flow variations;
- The velocity profile is considered pseudo-uniform in each section, assuming the values of momentum and Coriolis coefficients constant are equal to one;
- The behaviour of the pipe material is considered linear elastic;
- Head-losses are calculated by uniform flow friction formula, which is used in steady flow.

The differential Equations (1) and (2) can be simplified into a hyperbolic system of equations [36,39]. These equations can be presented as a matrix (4):

$$\frac{\partial U}{\partial t} + \frac{\partial F(U)}{\partial x} = D(U) \tag{4}$$

being:

$$U = \begin{bmatrix} H \\ Q \end{bmatrix}; F(U) = \begin{bmatrix} \frac{c^2}{gA} & Q \\ gA & H \end{bmatrix}; D(U) = \begin{bmatrix} 0 \\ \frac{-JgA}{Q^2}Q|Q| \end{bmatrix} \tag{5}$$

where J is the hydraulic gradient.

The solution of these equations is obtained through a discretized time interval for each time step 'Δt' at a specific point of the pipe for each 'Δx', fulfilling the Courant condition ($C_r = 1$) (6):

$$\frac{\Delta x}{\Delta t} = a \tag{6}$$

The differential Equation (4) can be transformed into linear algebraic equations, obtaining Equations (7) and (8). The application of these equations is denominated the "Method of Characteristics" (MOC).

$$C^+ : H_i^{n+1} - H_{i-1}^n + \frac{A}{c}\left(V_i^{n+1} - V_{i-1}^n\right) + \frac{f_{i-1}^n \Delta x}{D}V_{i-1}^n|V_{i-1}^n| = 0 \tag{7}$$

$$C^- : H_i^{n+1} - H_{i-1}^n - \frac{A}{c}\left(V_i^{n+1} - V_{i-1}^n\right) - \frac{f_{i-1}^n \Delta x}{D}V_{i-1}^n|V_{i-1}^n| = 0 \tag{8}$$

where H_i^{n+1} is the piezometric head in m w.c. at pipe section "i" and time instant "$n + 1$"; V_i^{n+1} is the velocity in m/s at pipe section "i" and time instant "$n + 1$"; where H_{i-1}^n is the piezometric head in mw.c. at pipe section "$i - 1$" and time instant "n"; V_{i-1}^n is the velocity in m/s at pipe section "$i - 1$" and time instant "n"; f_{i-1}^n is the friction factor in the section "$i - 1$" at time instant "n".

2.2. Control Valves

The valves are system components, which are responsible for changing the flow when its opening degree changes. Any operation in a valve modifies the opening degree and varies the loss coefficient of the valve causing a flow variation in the system, being one of the origin for hydraulic transients events. The closure time as well as the valve type influence the type of water hammer (i.e., fast or slow manoeuvers) for a system characterized by its diameter, length and pipe material.

For any manoeuvre, the loss coefficient of the valve is function of the opening degree [40] and in a simplistic characterization, the behaviour of the valve can be defined by the Equation (9):

$$Q(t) = Q_0\left(1 - \frac{t}{T_c}\right)^b \tag{9}$$

Figure 1 shows different closures as function of b exponent. If the exponent is one, the closure law is linear and the variation of the flow loss coefficient is continuous. When the exponent is less than one, the variation of the flow loss coefficient is higher at the end of the closure time (e.g., diaphragm valve-Figure 1a(f)). This significant difference in the loss coefficients (Kv) for different opening degrees will change the effective time closure defined in the Equation (10).

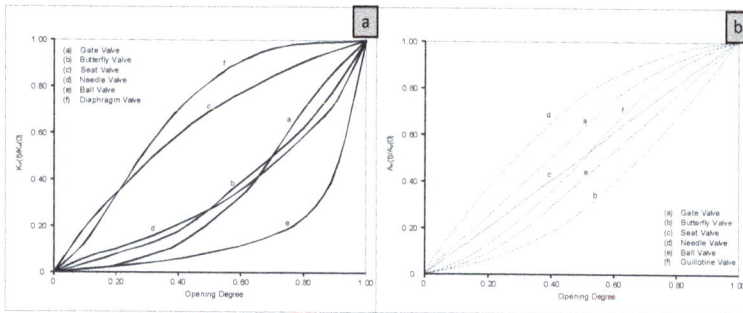

Figure 1. Valves manoeuvres. (**a**) type of loss coefficients and (**b**) opening cross area depending on the type of valve and the opening degree.

If the exponent is greater than one, the closure is higher at the beginning of the manoeuvre (e.g., butterfly valve), can cause higher overpressures since the main closure occurs when the velocity of the fluid is greater [41,42]. Although the closure law depends on the opening degree, by knowing the ratio of the free area as a function of the opening degree, the type of valve has great significance in the generated transient in a pipe system.

The duration of the valve manoeuvre, the diameter, the type of closure law (linear or non-linear) and the actuator type will influence the shape and values of the piezometric line envelopes. The effective time closure (T_{ef}) is the real time of valve closure (shorter than the total time (T_C)), which can induce high discharge reduction, responsible for the extreme water hammer phenomenon (as presented in Figure 2). Equation (10) mathematically defines the effective time closure based on the tangent to the point of the curvature in which dq/dt is highest:

$$T_{ef} = \frac{\Delta Q}{\left(\frac{dq}{dt}\right)_{max}} \tag{10}$$

where ΔQ is the discharge variation in the hydraulic system, q is the ratio Q/Q_o (relative discharge value) and Q_o is the discharge for total opening.

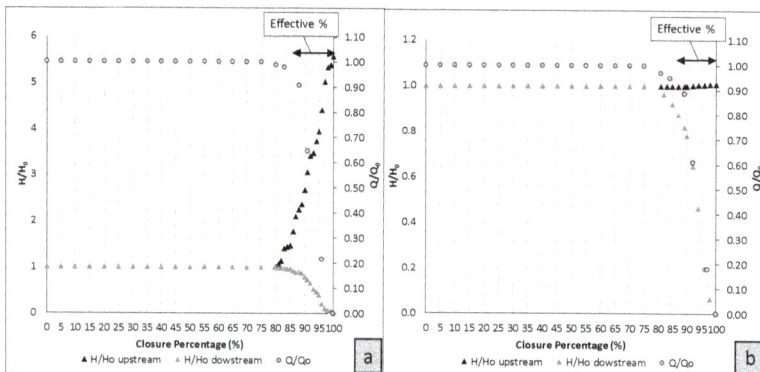

Figure 2. Comparison between effective closure and total closure of a ball valve: H/H_0 (upstream and downstream) variation and Q/Q_0. (**a**) turbulent flow (Re = 100,000) and (**b**) laminar flow (Re = 1000).

2.3. Damping Effects

Water hammer analysis usually focuses on the estimation of the extreme pressures associated with the valve manoeuvre, the pump trip-off or the turbine shutdown or start-up. The correct prediction of the pressure wave propagation, in particular, the damping effect, is not always properly accounted for. The latter will influence the system re-operation, the model calibration and the dynamic behaviour of the system response. Currently, transient solvers commercially available are not able to predict the observed damping pressure surge in real systems.

A new simplified approach of surge damping is presented considering the pressure peak damping in time. This damping can be a combined effect of the non-elastic behaviour of the pipe-wall and the unsteady friction effect, depending essentially upon the pipe material [43]. This technique aims at the characterization of energy dissipation through the variation of the extreme piezometric head over time [6,43–45].

In a rigid pipe with an elastic behaviour, the energy dissipation of the system over time for a rough turbulent flow, in a dimensionless form, varies with h^2 (due to almost exclusively friction effects). Based on the well-known upsurge given by the Joukowsky formulation through Equation (11):

$$\Delta H_j = \frac{cQ}{gS} \tag{11}$$

where c is the celerity wave in m/s; Q is the flow in m^3/s; g is gravity constant in m/s^2 and S is the section of the pipeline (m^2), the time head variation ($h = \frac{H}{\Delta H_j}$) can be obtained according to Equation (12):

$$h = \frac{1}{\frac{1}{h_0} + K\Delta h_0(\tau - \tau_0)} \tag{12}$$

assuming $\tau = \frac{t}{\frac{2L}{c}}$, being h_0 the dimensionless head at initial time, $\tau_0 = \frac{t_0}{\frac{2L}{c}}$, and t_0 the time for the first pressure peak where the head is maximum.

According to the same type of analysis, in a plastic pipe with a non-elastic behaviour (e.g., PVC, HDPE), the pipe-wall retarded-behaviour is mainly responsible for the pressure damping. Thus, the energy dissipation can adequately be reproduced with mathematic transformations using Equation (13) [43,44]:

$$h = h_0 e^{-K\Delta h_0(\tau - \tau_0)} \tag{13}$$

This equation is in accordance with the typical behaviour of a viscoelastic solid. For systems with combined effects (i.e., elastic and plastic), the surge damping can be evaluated by the combination of both former effects through Equation (14) [43,44]

$$h = \frac{1}{\left(\frac{K_{elas}}{K_{plas}} + \frac{1}{h_0}\right)e^{K_{plas}\Delta h_0(\tau - \tau_0)} - \frac{K_{elas}}{K_{plas}}} \tag{14}$$

where K_{plas} and K_{elas} are decay coefficients for the plastic and elastic effects, respectively.

2.4. Runaway Conditions

The specific rotational speed (n_s) given by Equation (15):

$$n_s = n\frac{P^{1/2}}{H^{5/4}} \tag{15}$$

where n_s is the specific speed of the machine in (m, kW); n is the rotational speed of the machine in rpm; P is the power in the shaft, which is measured in (kW); and H is the recovered head in (m w.c.), a characteristic parameter describing the runner shape and its associated dynamic behavior.

The flow drops with the transient overspeed in reaction turbines with low specific speed. Conversely, the transient discharge tends to increase for turbines with high specific speed [6,35,44–49].

The flow across a runner is characterized by three types of velocities: absolute velocity of the water (V) with the direction imposed by the guide vane blade, relative velocity (W) through the runner and tangential velocity (C) of the runner.

If a uniform velocity distribution is assumed at inlet (Section 1) and outlet (Section 2) of a runner, the application of Euler's theorem enables us to obtain the relation between the motor binary and the momentum moment between Sections 1 and 2 by Equation (16):

$$BH = \rho Q(r_1 V_1 \cos\alpha_1 - r_2 V_2 \cos\alpha_2) \tag{16}$$

where α and r are the angle and radius, respectively (Figure 3).

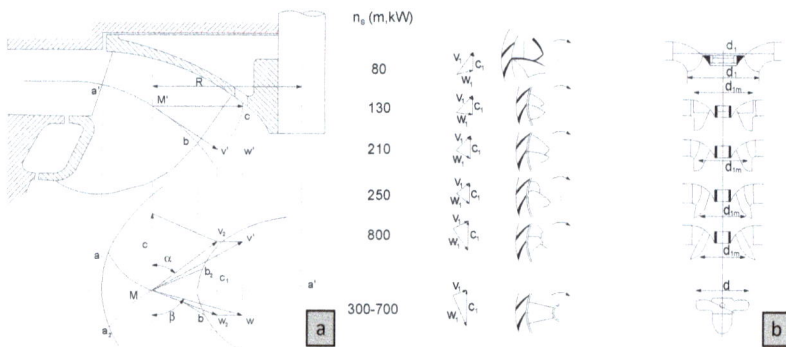

Figure 3. Velocity components across a reaction turbine runner. (**a**) scheme of an impeller and (**b**) velocity vectors (adapted from [6,35,44–50]).

The output power in the shaft is defined by Equation (17).

$$P = BH \cdot \omega \tag{17}$$

where BH is the hydraulic torque in Nm and P the output power in W.

The velocity components (Figure 3) at the inlet and outlet of a runner allow us to obtain the ratio between the flow discharge under runaway conditions (Q_{RW}) and the discharge for initial conditions (Q_0), which lean towards a linear increase with the rise of the specific speed (Figure 4) [6,35,45].

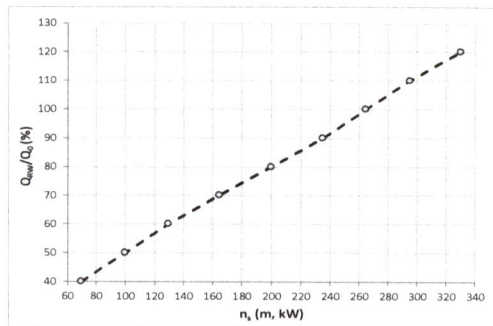

Figure 4. Overspeed effect on the discharge variation (Q_{RW}/Q_0) of reaction turbines (adapted from [6,35]).

Furthermore, the variations of the ratio as a function of N/N_{BEP} for constant values of h (H/H_{BEP}) are shown in Figure 5 for radial and axial conventional turbines. Q/Q_{BEP} are based on Suter parameters which are in accordance with the dynamic behaviour associated with the runner shape [35].

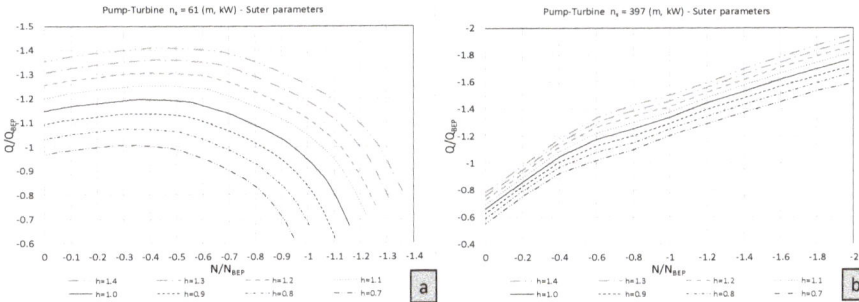

Figure 5. Q/Q_{BEP} as function of N/N_{BEP} and h for (**a**) radial and (**b**) axial machine (adapted from [47–49]).

3. Results and Discussion

3.1. Experiments and Simulations

The identification of turbomachines that can be used in pressurized systems are of action or reaction types. In the reaction machines, the hydraulic power is transmitted to the axis of the machine by varying the pressure flow between the inlet and outlet of the impeller, which depends on the specific speed of the machine (e.g., Francis, propeller and Kaplan). In action turbines, the energy exchange (hydraulic to mechanical) is carried out at atmospheric pressure, and the hydraulic power is due to the kinetic energy of the flow (e.g., Pelton and Turgo).

Experimental tests were carried out in the CERIS-Hydraulic Lab of Instituto Superior Técnico from the University of Lisbon for a radial and an axial reaction machine with small size (Figure 6a). A small pressurized system was installed in order to develop the experimental test. The facility scheme (Figure 6b) is composed of: (1) a reservoir to collect and supply the water looped facility; (2) one pump to recirculate the flow; (3) an air vessel to guarantee the uniform pressure in the pipe gravity system; (4) 100 m of HDPE pipeline or 25 m of PVC pipe for experiments with radial or axial PAT, respectively; (5) a radial or axial PAT depending on the selected machine. In both cases, the discharge was measured by an electromagnetic flowmeter; the pressure was registered by pressure transducers, through the Picoscope data acquisition system; the power was measured by a Wattmeter which was connected to the generator; and the rotational speed was measured by a frequency meter.

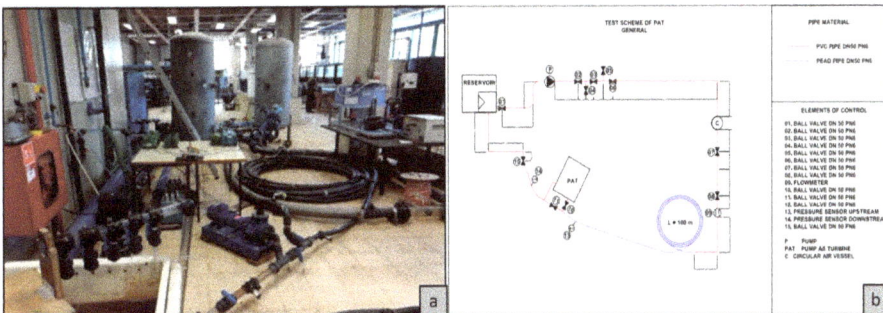

Figure 6. Lab set-up. (**a**) experimental facility at IST; (**b**) scheme of the facility [6].

The used machines were a radial pump working as turbine (PAT), with a rotational specific speed of 51 rpm (in m, kW); and an axial one, with a rotational specific speed of 283 rpm (in m, kW) (Figure 7). Each machine was tested on different hydraulic circuits according to the available facilities. The radial machine scheme was composed of a reservoir to stabilize the flow; a pump to recirculate the flow; an air-vessel tank to control and stabilize the system pressure, which had a 1 m^3 capacity; an electromagnetic flowmeter; one hundred meters of high density polyethylene (HDPE) pipe, with 50 mm nominal diameter; a PAT which is connected downstream of the HDPE loop pipe; and a ball valve located downstream of the PAT.

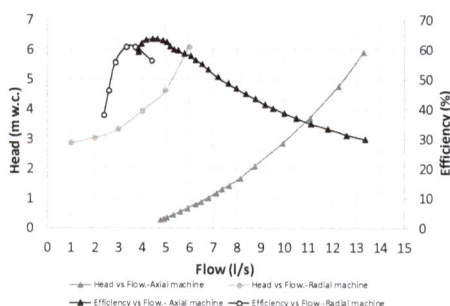

Figure 7. Experimental Head and Efficiency curves of axial (N_0 = 750 rpm) and radial (N_0 = 1020 rpm) machines.

The ball valve was connected to the reservoir by a PVC pipe. The pump and air vessel were joined by a PVC pipe of 7.40 m of length and 50 mm of nominal diameter. The air vessel and flowmeter were joined by another rigid PVC pipe 1.80 m long. Two pressure sensors were installed upstream and downstream of the PAT to estimate the net head.

For the axial machine, the scheme was similar to the previous one. The facility was composed of a reservoir, a pump to recirculate the flow, an air vessel tank to maintain a quasi-uniform pressure (the capacity of this tank is 1 m^3), an electromagnetic flowmeter to measure the flow and the axial machine, which is followed by a butterfly valve to isolate the facility. The pump and the air vessel were joined by a steel pipe with a length of 3.50 m and diameter of 80 mm. The axial machine and the butterfly valve were connected by a pipe, which is composed of PVC (4.90 m and 110 mm of diameter) and a steel pipe (4.50 m and 80 mm of diameter). The butterfly valve and the reservoir were connected by a steel pipe, 2 m long, with a diameter equal to 80 mm. Two pressure sensors were installed upstream and downstream of the axial machine.

These hydraulic systems (Figure 8) were simulated by Allievi software [51] according to the system characteristics in each facility, previously described. The inner diameter and the wave speed are shown in Table 1 according to the pipe material. During the simulation process, all singularities were verified and the friction losses along the pipe system were defined, adopting the following procedure with excellent results: (1) a model calibration for the friction factor (pipe roughness) and for the singular head losses was done, considering different singularities, such as ball valves, inlet and outlet of the air-vessel, elbows, bifurcations and valve connections; this setting under steady state conditions that allowed us to make small refinements after comparisons with the experiments; (2) due to the system characteristics, it was also possible to fit the damping effect, as well as the phase shift of the pressure waves during unsteady state conditions; based on the authors experience, it was concluded that in viscoelastic pipes and for slow manoeuvres, the unsteady friction has no significance in terms of the damping and the shape of the pressure wave propagation. The dynamic mathematical model considers the internal friction losses to analyse the presence of a PAT in a water distribution network with suitable results [46,47].

The simulated flow and the head in the pump, as well as the pressure in the air vessel, were calibrated in each developed simulation (i.e., for radial and axial machines) according to the registered experimental data for each test type. When the radial machine was tested, the flow values oscillated between 1 and 7 l/s, the upstream pressure between 15 and 30 m, and the head drop between 3 and 10 m. In the axial machine, the tested flow varied between 5 and 14.1 l/s, the upstream pressure between 10 and 20 m and the head drop between 0.25 and 7 m, as previously designed [20,50].

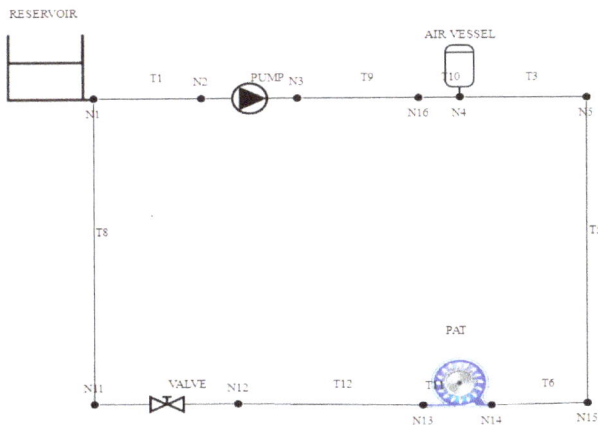

Figure 8. Simulation scheme used in Allievi model.

Table 1. Basic parameters for the simulation.

Material	Inner Diameter (m)	Roughness (mm)	Wave Speed (m/s)
HDPE	0.044	0.2	280
PVC	0.110	1.2	385
Rigid PVC	0.047	0.2	527
Steel	0.068	2	1345

3.2. Control Valve Closure and PAT Trip-Off

This section shows the flow and the rotational speed variation when a fast shutdown was carried out downstream of the PAT. Figure 9 shows four tests with different initial flow values in the radial machine and three tests for the axial one. The values rapidly varied from the nominal values (flow and rotational speed) to zero. The closure time is around two seconds in all considered manoeuvres.

According to the installed systems, the model was implemented in Allievi software as the scheme presented in Figure 8 shows. The software is a computational model that enables us to analyse water systems (pressurized and open channel flows) under steady and unsteady conditions. The developed model was calibrated to consider the damping effects that were associated with the characteristic parameters of the system, as well as the type of hydraulic machines.

Comparisons between experimental and simulated pressure values (upstream and downstream) presented adequate fitting. In Figure 10, the experimental overpressure in the radial machine was 69.85 m w.c. while the simulated overpressure was 70.52 m w.c. The result in the first depression wave was similar, where the minimum experimental value was 24.54 m w.c. and the simulated was 19.73 m w.c. The axial machine presented values of 46.23 m w.c. (experimental) and 49.24 m w.c. (simulated). The minimum experimental depression value was 36.54 m w.c. while the simulated value was 33.63 m w.c. These results showed the dynamic behaviour of the radial and axial machines when a downstream induced transient attained the turbine runner. The transient wave passed through the runners and the pressure variation upstream and downstream was essentially in phase.

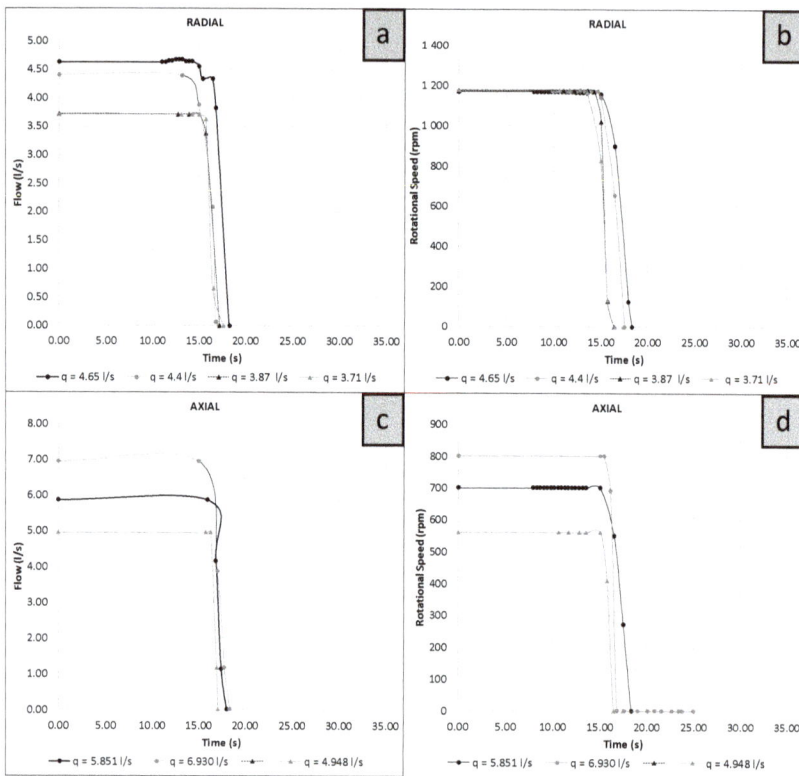

Figure 9. Experimental data recorded for the fast closure of the downstream control valve in radial and axial turbine machines. (**a**) flow for radial machine (**b**) rotational speed for radial machine (**c**) flow for axial machine (**d**) rotational speed for axial machine.

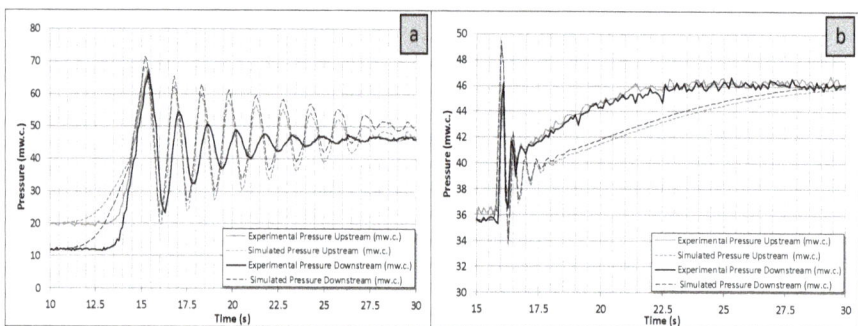

Figure 10. Experimental and simulated pressure values along time in a fast closure manoeuvre ($t = 2$ s): (**a**) radial and (**b**) axial machine.

3.3. Control Valve Opening and PAT Start-Up

The flow, the rotational speed, and the pressure values (upstream and downstream) were recorded over time (Figures 11 and 12). Figure 11 shows the flow and speed variation for a fast opening of the downstream control valve for each system. Some variations can be observed, where the rotational speed of the machine reached 2235 rpm. This value was double the nominal rotational speed of the machine for 4.65 *l/s* in the radial machine. The reached value was 1500 rpm for the axial machine, when the nominal flow and the nominal rotational speed were 6.93 *l/s*.

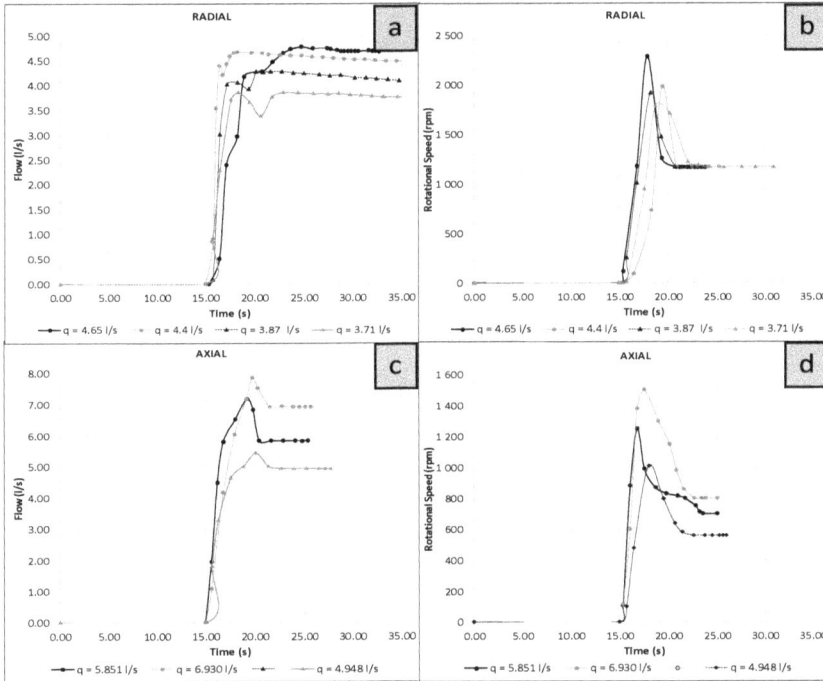

Figure 11. Experimental data recorded flow and rotational speed for the fast opening of the downstream control valve in turbine machines. (**a**) flow for radial machine (**b**) rotational speed for radial machine (**c**) flow for axial machine (**d**) rotational speed for axial machine.

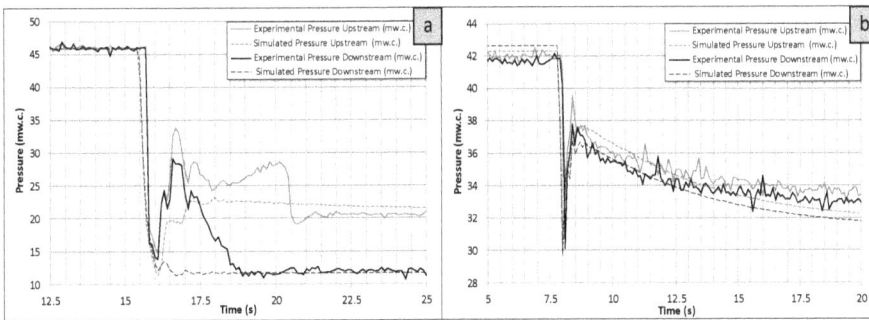

Figure 12. Experimental data and simulation for pressure variation due to a fast opening downstream control valve of (**a**) radial and (**b**) axial machine.

The trend in the valve opening for PAT start-up was similar in all cases: firstly, the machine increased the rotational speed upper its nominal value. When the overspeed was reached for flow value near 4.00 *l/s* (in radial machine) and 7.86 *l/s* (in axial machine), the rotational speed decreased to the nominal one. In this time, the flow attained the nominal flow with the maximum valve opening degree. A downsurge wave in both machines (radial and axial) was observed with the valve opening. This depression depended on the flow and the opening time. This value was also simulated with Allievi, achieving quite accurate results (Figure 12). The root mean square error (RMSE) for each simulation with Allievi was determined. The average RMSE obtained in the first phase of the water hammer phenomenon (i.e., worst value) was 2.37%, and the standard deviation was 3.47%. When the maximum downsurge and upsurge were compared to the experiments, the maximum error was 1.26%.

3.4. Overspeed Effect in PATs

Some interesting conclusions can be drawn for both types of runners. Figure 13 presents the obtained values of flow, rotational speed, and pressure (upstream and downstream) for the overspeed conditions. The flow value decreased over time in all tests induced by the runner shape associated with the low specific speed value as previously mentioned from Figures 3–5. This decrease of the flow was related to an increase in the rotational speed, with the minimum flow attained when the runaway conditions were reached.

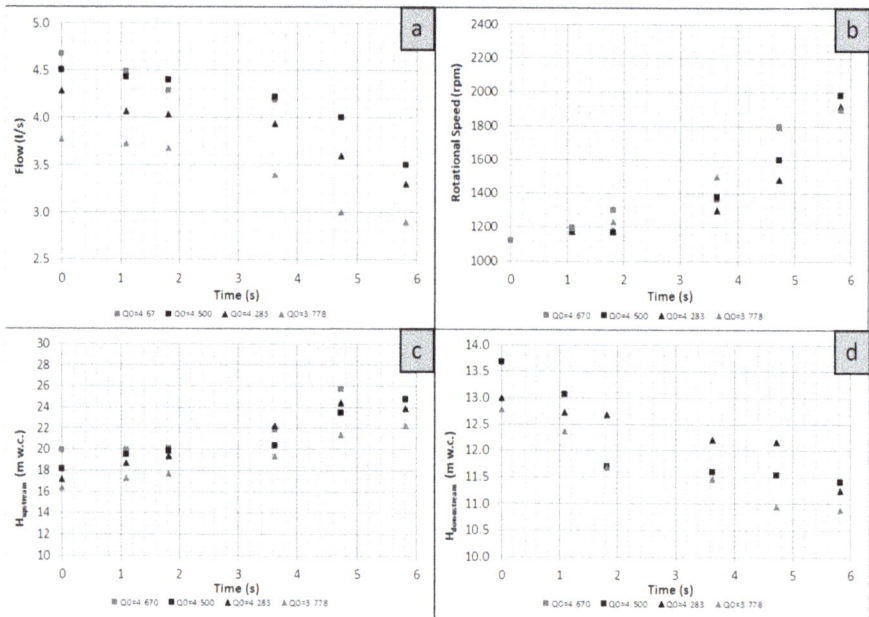

Figure 13. Experimental data of flow, rotational speed, and upstream and downstream head in the radial machine under the overspeed effect. (**a**) flow for radial machine (**b**) rotational speed for radial machine (**c**) flow for axial machine (**d**) rotational speed for axial machine.

Regarding the pressure variation, some issues can be observed: the upstream pressure value increased, resulting in the maximum pressure when the machine reaches the runaway conditions and consequently, the downstream pressure decreased (Figure 13). Therefore, the radial runner induced a flow cut effect under overspeed conditions.

The experimental data under runaway conditions can be expressed by different parameters: the discharge flow, the pressure and the rotational speed for total opening valve degree (Q_0, H_0, N_0). Furthermore, the experimental results can be associated with the values of the best efficiency point of the machine in turbine mode (Q_{BEP}, H_{BEP}, N_{BEP}). These variations are shown in Figure 14. If Q_{RW}/Q_0 versus N_{RW}/N_0 (the subscripts 'RW' indicates runaway conditions) is observed, the values were almost constant for all experimental data denoting a typical characteristic of the radial machine. In this case, the ratio Q_{RW}/Q_0 is near 0.514; therefore, there was a flow reduction of around 50%. This value is close to the presented value in Figure 4 that shows the characteristic of the radial machine under the overspeed effect. Similar conclusions can be obtained if the upstream and downstream pressures are analysed in the axial machine. In this case, the values were near 1.40 and 0.85, inducing an upsurge and a downsurge upstream and downstream, respectively, of the machine. If the values are compared with the best efficiency point of the radial machine, under the overspeed effect, the flow decreased for a constant value of h ($h = H/H_{BEP}$).

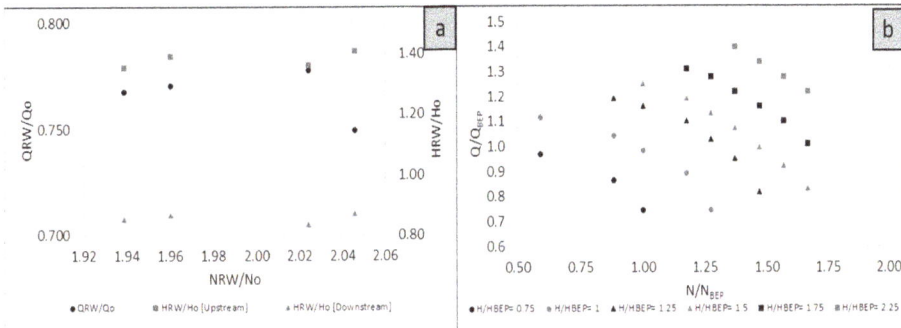

Figure 14. (**a**) Q_{RW}/Q_0 and H_{RW}/H_0 as a function of N_{RW}/N_0 and (**b**) Q/Q_{BEP} as a function of N/N_{BEP} and H/H_{BEP} for the radial machine.

Converse results were obtained when the experimental data were analyzed for the axial machine (Figure 15). The flow rise over time as the rotational speed increased until to reach the runaway value. In these cases, the upstream and downstream pressures remained almost constant over time.

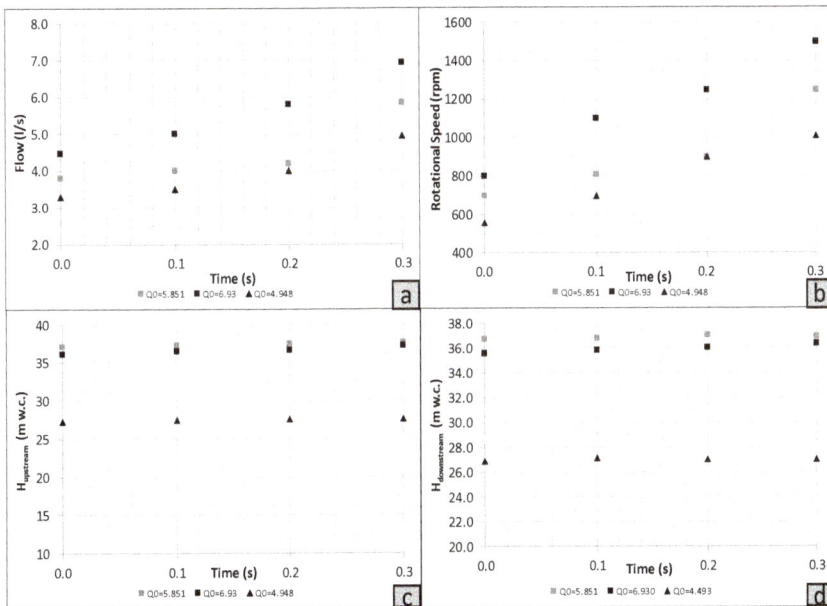

Figure 15. Experimental data of (**a**) flow, (**b**) rotational speed, and (**c**) the upstream and (**d**) downstream head in the axial machine under the overspeed effect.

As for the radial machine, Figure 16 shows the experimental data for the discharge flow, pressure, and the rotational speed variation for the total valve opening (Q_0, H_0, N_0) during the overspeed conditions of the axial machine. In this case, the ratio Q_{RW}/Q_0 showed an increase in flow. This value is higher than the obtained value using Figure 4, for n_s of 280 rpm (in m, kW).

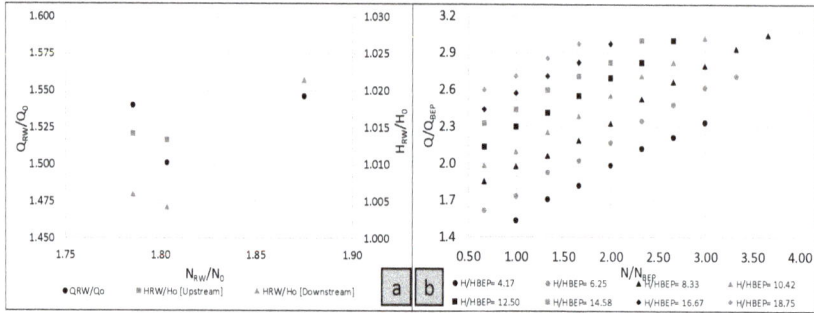

Figure 16. (**a**) Q_{RW}/Q_0 and H_{RW}/H_0 as a function of N_{RW}/N_0 and (**b**) Q/Q_{BEP} as a function of N/N_{BEP} and $h = H/H_{BEP}$ for the axial machine.

The experimental data were also correlated with the values of the best efficiency point (Q_{BEP}, H_{BEP}, N_{BEP}). The results contrasted with those obtained for the radial machine. Under a constant value of h ($h = H/H_{BEP}$), the flow increased when the rotational speed increased. Figure 16 shows all analysed cases considering a constant h value and they present the same tendency.

4. Conclusions

Based on the authors' experience and laboratory tests, a large number of criteria to deal with the design and hydraulic transients of micro hydropower systems are addressed. The type of analysis will be influenced by the design stage and the complexity of each system. Hence, based on each hydraulic system characteristic, for the most predictable manoeuvres, the designers will be able to define exploitation rules according to expected safety levels. In fact, convenient operational rules need to be specified in order to control the maximum and minimum transient pressures. These specifications will mainly depend on the following factors:

- the characteristics of the pipe system to be protected; in fact, these characteristics based on the head loss and inertia of the water column can adversely modify the system behaviour and the same valve closure time can induce a slow or a rapid flow change;
- the intrinsic characteristics of the valve: a butterfly valve (e.g., for medium heads) and a spherical valve (e.g., for high heads) have different effects on the dynamic flow response for the same closure law;
- since PATs have no guide vane, the flow control is made through valves where the closure and opening laws are crucial in the safety system conditions, such as the type of the valve actuator;
- based on the characteristics of the pump such as turbine machine (i.e., radial or axial), different dynamic behaviour will be associated with:

  ○ the small inertia of the rotating masses induces a fast overspeed effect under runaway conditions imposed by a full load rejection.
  ○ the overspeed effects provoke flow variations (i.e., flow reduction in low n_s machines and flow increasing in the high n_s machines) and pressure variations that can propagate upsurges upstream of a radial machine and downsurges downstream of it, in contrast to axial machines (downsurges upstream and upsurges downstream).

As a novelty, the manuscript analyses the unsteady flow behaviour in real PATs system. This analysis is based on data obtained during an intensive experimental campaign, and important information is presented and utilized for some specific PAT transient state conditions, using a computational tool that was previously calibrated.

This analysis required a mathematical transformation of available data of pumps (based on experiments especially developed for this study) into characteristic curves of discharge variation, Q/Q_{BEP}, with the rotating speed, N/N_{BEP} (Figures 14 and 16, for radial and axial machines, respectively). This procedure facilitates understanding of the dynamic pump as turbine behaviour under unsteady conditions.

The feasibility of pumps operating as turbines was proved, based on typical performed control analyses. The dynamic behaviour of those machines presents similarities to the classical reaction turbines, regarding the flow variation due to the runner type, generally characterized by its specific rotational speed (n_s) [35,45,47,51] apart from the associated scale effects.

These PAT solutions can be adopted instead of energy dissipation devices in conveyance pipe systems with excess available energy at some pipe sections. Therefore, the use of reverse pumps in drinking, irrigation and sewage or drainage water systems can be a rather interesting solution in some cases, taking the advantage of the available head that in another way would be dissipated.

Author Contributions: The author Helena M. Ramos has contributed with the idea, wrote part of the document, was in the revision of the document and supervised the whole research. Modesto Pérez-Sánchez did the experiments and wrote part of the document. P. Amparo López-Jiménez suggested guides towards the developed analyses.

Funding: The authors wish to thank to the project REDAWN (Reducing Energy Dependency in Atlantic Area Water Networks) EAPA_198/2016 from INTERREG ATLANTIC AREA PROGRAMME 2014–2020 and CERIS (CEHIDRO-IST). This research was developed in the research stay of the first author in the hydraulic lab of CERIS-IST in January 2018 called "MAXIMIZATION OF THE GLOBAL EFFICIENCY IN PATs IN LABORATORY FACILITY".

Conflicts of Interest: The authors have declared that no competing interests exist.

References

1. Kougias, I.; Patsialis, T.; Zafirakou, A.; Theodossiou, N. Exploring the potential of energy recovery using micro hydropower systems in water supply systems. *Water Util. J.* **2014**, *7*, 25–33.

2. Nogueira, M.; Perrella, J. Energy and hydraulic efficiency in conventional water supply systems. *Renew. Sustain. Energy Rev.* **2014**, *30*, 701–714. [CrossRef]

3. Moreno, M.; Córcoles, J.; Tarjuelo, J.; Ortega, J. Energy efficiency of pressurised irrigation networks managed on-demand and under a rotation schedule. *Biosyst. Eng.* **2010**, *107*, 349–363. [CrossRef]

4. Jiménez-Bello, M.A.; Royuela, A.; Manzano, J.; Prats, A.G.; Martínez-Alzamora, F. Methodology to improve water and energy use by proper irrigation scheduling in pressurised networks. *Agric. Water Manag.* **2015**, *149*, 91–101. [CrossRef]

5. Cabrera, E.; Cabrera, E., Jr.; Cobacho, R.; Soriano, J. Towards an Energy Labelling of Pressurized Water Networks. *Procedia Eng.* **2014**, *70*, 209–217. [CrossRef]

6. Carravetta, A.; Houreh, S.D.; Ramos, H.M. *Pumps as Turbines: Fundamentals and Applications*; Springer International Publishing: Cham, Switzerland, 2018; p. 218. ISBN 978-3-319-67507-7.

7. Abbott, M.; Cohen, B. Productivity and efficiency in the water industry. *Util. Policy* **2009**, *17*, 233–244. [CrossRef]

8. Araujo, L.; Ramos, H.; Coelho, S. Pressure Control for Leakage Minimisation in Water Distribution Systems Management. *Water Resour. Manag.* **2006**, *20*, 133–149. [CrossRef]

9. Dannier, A.; Del Pizzo, A.; Giugni, M.; Fontana, N.; Marini, G.; Proto, D. Efficiency evaluation of a micro-generation system for energy recovery in water distribution networks. In Proceedings of the 2015 International Conference on Clean Electrical Power (ICCEP), Taormina, Italy, 16–18 June 2015; pp. 689–694.

10. Giugni, M.; Fontana, N.; Ranucci, A. Optimal Location of PRVs and Turbines in Water Distribution Systems. *J. Water Resour. Plan. Manag.* **2014**, *140*, 06014004. [CrossRef]

11. Ramos, H.; Borga, A. Pumps as turbines: An unconventional solution to energy production. *Urban Water*
     1999, *1*, 261–263. [CrossRef]
12. Pérez-Sánchez, M.; Sánchez-Romero, F.; Ramos, H.; López-Jiménez, P.A. Energy Recovery in Existing Water
     Networks: Towards Greater Sustainability. *Water* **2017**, *9*, 97. [CrossRef]
13. Senior, J.; Saenger, N.; Müller, G. New hydropower converters for very low-head differences. *J. Hydraul. Res.*
     2010, *48*, 703–714. [CrossRef]
14. Razan, J.I.; Islam, R.S.; Hasan, R.; Hasan, S.; Islam, F. A Comprehensive Study of Micro-Hydropower Plant
     and Its Potential in Bangladesh. *ISRN Renew. Energy* **2012**, *2012*, 635396. [CrossRef]
15. Elbatran, A.H.; Yaakob, O.B.; Ahmed, Y.M.; Shabara, H.M. Operation, performance and economic analysis of
     low head micro-hydropower turbines for rural and remote areas: A review. *Renew. Sustain. Energy Rev.* **2015**,
     43, 40–50. [CrossRef]
16. Nourbakhsh, A.; Jahangiri, G. Inexpensive small hydropower stations for small areas of developing countries.
     In Proceedings of the Conference on Advanced in Planning-Design and Management of Irrigation Systems
     as Related to Sustainable Land Use, Louvain, Belgium, 14–17 September 1992; pp. 313–319.
17. Simão, M.; Ramos, H.M. Hydrodynamic and performance of low power turbines: Conception, modelling
     and experimental tests. *Int. J. Energy Environ.* **2010**, *1*, 431–444.
18. Arriaga, M. Pump as turbine—A pico-hydro alternative in Lao People's Democratic Republic. *Renew. Energy*
     2010, *35*, 1109–1115. [CrossRef]
19. Pérez-Sánchez, M.; López Jiménez, P.A.; Ramos, H.M. Modified Affinity Laws in Hydraulic Machines
     towards the Best Efficiency Line. *Water Resour. Manag.* **2018**, *3*, 829–844. [CrossRef]
20. Ramos, H.M.; Borga, A.; Simão, M. New design solutions for low-power energy production in water pipe
     systems. *Water Sci. Eng.* **2009**, *2*, 69–84.
21. Carravetta, A.; Del Giudice, G.; Fecarotta, O.; Ramos, H.M. Energy Recovery in Water Systems by PATs:
     A Comparisons among the Different Installation Schemes. *Procedia Eng.* **2014**, *70*, 275–284. [CrossRef]
22. Caxaria, G.; de Mesquita e Sousa, D.; Ramos, H.M. Small Scale Hydropower: Generator Analysis and
     Optimization for Water Supply Systems. 2011, p. 1386. Available online: http://www.ep.liu.se/ecp_article/
     index.en.aspx?issue=57;vol=6;article=2 (accessed on 12 March 2017).
23. Butera, I.; Balestra, R. Estimation of the hydropower potential of irrigation networks. *Renew. Sustain. Energy Rev.*
     2015, *48*, 140–151. [CrossRef]
24. Carravetta, A.; Fecarotta, O.; Del Giudice, G.; Ramos, H. PAT Design Strategy for Energy Recovery in Water
     Distribution Networks by Electrical Regulation. *Energies* **2013**, *6*, 411–424. [CrossRef]
25. Fecarotta, O.; Aricò, C.; Carravetta, A.; Martino, R.; Ramos, H.M. Hydropower Potential in Water Distribution
     Networks: Pressure Control by PATs. *Water Resour. Manag.* **2014**, *29*, 699–714. [CrossRef]
26. Fecarotta, O.; Carravetta, A.; Ramos, H.M.; Martino, R. An improved affinity model to enhance variable
     operating strategy for pumps used as turbines. *J. Hydraul. Res.* **2016**, *54*, 332–341. [CrossRef]
27. Sitzenfrei, R.; Berger, D.; Rauch, W. Design and optimization of small hydropower systems in water
     distribution networks under consideration of rehabilitation measures. *Urban Water J.* **2015**, *12*, 1–9. [CrossRef]
28. De Marchis, M.; Milici, B.; Volpe, R.; Messineo, A. Energy Saving in Water Distribution Network through
     Pump as Turbine Generators: Economic and Environmental Analysis. *Energies* **2016**, *9*, 877. [CrossRef]
29. Samora, I.; Manso, P.; Franca, M.; Schleiss, A.; Ramos, H. Energy Recovery Using Micro-Hydropower
     Technology in Water Supply Systems: The Case Study of the City of Fribourg. *Water* **2016**, *8*, 344. [CrossRef]
30. Pérez-Sánchez, M.; Sánchez-Romero, F.; Ramos, H.; López-Jiménez, P.A. Modeling Irrigation Networks for
     the Quantification of Potential Energy Recovering: A Case Study. *Water* **2016**, *8*, 234. [CrossRef]
31. Corcoran, L.; McNabola, A.; Coughlan, P. Predicting and quantifying the effect of variations in long-term
     water demand on micro-hydropower energy recovery in water supply networks. *Urban Water J.* **2016**, *9*, 1–9.
     [CrossRef]
32. Pérez-Sánchez, M.; Sánchez-Romero, F.J.; Ramos, H.M.; López Jiménez, P.A. Optimization Strategy for
     Improving the Energy Efficiency of Irrigation Systems by Micro Hydropower: Practical Application. *Water*
     2017, *9*, 799. [CrossRef]
33. Imbernón, J.A.; Usquin, B. Sistemas de generación hidráulica. Una nueva forma de entender la energía.
     In Proceedings of the II Congreso Smart Grid, Madrid, Spain, 27–28 October 2014.

34. McNabola, A.; Coughlan, P.; Corcoran, L.; Power, C.; Prysor, A.; Harris, I.; Gallagher, J.; Styles, D. Energy recovery in the water industry using micro-hydropower: An opportunity to improve sustainability. *Water Policy* **2014**, *16*, 168–183. [CrossRef]

35. Ramos, H. Simulation and Control of Hydrotransients at Small Hydroelectric Power Plants. Ph.D. Thesis, IST, Lisbon, Portugal, December 1995.

36. White, F.M. *Fluid Mechanics*, 6th ed.; McGrau-Hill: New York, NY, USA, 2008.

37. Wylie, E.B.; Streeter, V.L. *Fluid Transients in Systems*; Prentice Hall: Englewood Cliffs, NI, USA, 1993.

38. Almeida, A.B.; Koelle, E. *Fluid Transients in Pipe Networks*; Computational Mechanics Publications, Elsevier Applied Science: Amsterdam, The Netherlands, 1992.

39. Chaudhry, M. *Applied Hydraulic Transients*, 2nd ed.; Springer-Verlag: New York, NY, USA, 1987.

40. Abreu, J.; Guarga, R.; Izquierdo, J. *Transitorios y Oscilaciones en Sistemas Hidráulicos a Presión*; Abreu, J., Guarga, R., Izquierdo, J., Eds.; U.D. Mecánica de Fluidos, Universidad Politécnica de Valencia: Valencia, Spain, 1995.

41. Iglesias-Rey, P.; Izquierdo, J.; Fuertes, V.; Martínez-Solano, F. *Modelación de Transitorios Hidráulicos Mediante Ordenador*; Grupo Mult.; Universidad Politécnica de Valencia: Valencia, Spain, 2004.

42. Subani, N.; Amin, N. Analysis of Water Hammer with Different Closing Valve Laws on Transient Flow of Hydrogen-Natural Gas Mixture. *Abstr. Appl. Anal.* **2015**, *2*, 12–19. [CrossRef]

43. Ramos, H.M.; Covas, D.; Borga, A.; Loureiro, D. Surge damping analysis in pipe systems: Modelling and experiments. *J. Hydraul. Res.* **2004**, *42*, 413–425. [CrossRef]

44. Ramos, H. Design concerns in pipe systems for safe operation. *Dam Eng.* **2003**, *14*, 5–30.

45. Ramos, H. *Guidelines for Design of Small Hydropower Plants*; Western Regional Energy Agency & Network (WREAN); Department of Economic Development (DED): Belfast, UK, 2000.

46. Ramos, H.; Almeida, A.B. Dynamic orifice model on water hammer analysis of high or medium heads of small hydropower schemes. *J. Hydraul. Res.* **2001**, *39*, 429–436. [CrossRef]

47. Ramos, H.; Almeida, A.B. Parametric Analysis of Water-Hammer Effects in Small Hydro Schemes. *J. Hydraul. Eng.* **2002**, *128*, 689–696. [CrossRef]

48. Ramos, H.M.; Simão, M.; Borga, A. Experiments and CFD Analyses for a New Reaction Microhydro Propeller with Five Blades. *J. Energy Eng.* **2013**, *139*, 109–117. [CrossRef]

49. Mataix, C. *Turbomáquinas Hidráulicas*; Universidad Pontificia Comillas: Madrid, Spain, 2009.

50. De Marchis, M.; Fontanazza, C.M.; Freni, G.; Messineo, A.; Milici, B.; Napoli, E.; Notaro, V.; Puleo, V.; Scopa, A. Energy recovery in water distribution networks. Implementation of pumps as turbine in a dynamic numerical model. *Procedia Eng.* **2014**, *70*, 439–448. [CrossRef]

51. ITA. Allievi, 2010. Available online: www.allievi.net (accessed on 17 July 2017).

Article

Topological Taxonomy of Water Distribution Networks

Carlo Giudicianni [1,2], **Armando Di Nardo** [1,2,3,*], **Michele Di Natale** [1,2], **Roberto Greco** [1,2], **Giovanni Francesco Santonastaso** [1,2] **and Antonio Scala** [3]

[1]   Dipartimento di Ingegneria, Universitá degli Studi della Campania Luigi Vanvitelli, via Roma 29, 81031 Aversa, Italy; carlo.giudicianni@unicampania.it (C.G.); michele.dinatale@unicampania.it (M.D.N.); roberto.greco@unicampania.it (R.G.); giovannifrancesco.santonastaso@unicampania.it (G.F.S.)

[2]   Action Group CTRL+SWAN of the European Innovation Partnership on Water, EU, via Roma 29, 81031 Aversa, Italy

[3]   Istituto Sistemi Complessi (Consiglio Nazionale delle Ricerche), via dei Taurini 19, 00185 Roma, Italy; antonio.scala.phys@gmail.com

*   Correspondence: armando.dinardo@unicampania.it; Tel.: +39-081-501-0202

Received: 13 February 2018; Accepted: 1 April 2018; Published: 8 April 2018

Abstract: Water Distribution Networks (WDNs) can be regarded as complex networks and modeled as graphs. In this paper, Complex Network Theory is applied to characterize the behavior of WDNs from a topological point of view, reviewing some basic metrics, exploring their fundamental properties and the relationship between them. The crucial aim is to understand and describe the topology of WDNs and their structural organization to provide a novel tool of analysis which could help to find new solutions to several arduous problems of WDNs. The aim is to understand the role of the topological structure in the WDNs functioning. The methodology is applied to 21 existing networks and 13 literature networks. The comparison highlights some topological peculiarities and the possibility to define a set of best design parameters for ex-novo WDNs that could also be used to build hypothetical benchmark networks retaining the typical structure of real WDNs. Two well-known types of network ((a) square grid; and (b) random graph) are used for comparison, aiming at defining a possible mathematical model for WDNs. Finally, the interplay between topology and some performance requirements of WDNs is discussed.

Keywords: water distribution network management; complex network theory; topological analysis; mathematical model

1. Introduction

Biological and chemical systems, brain neural networks, social interacting species, the Internet and the World Wide Web, and the multiple and interconnected infrastructures that provide several services to consumers in the cities are network shaped [1,2]. It seems that the efficiency of the systems largely depends on their ability to create (if they are natural) or to have (in the case of man-made structures) multiple links between the units. On the one hand, it ensures better performance and self-recovering function in the case of failure of an element thanks to a redundant structure. On the other hand, it makes it difficult to understand the principles of their functioning and behavior. In this regard, a suitable approach to capture the local and the global properties of network systems is to model them as graphs, where the nodes represent the units, and the links stand for the interactions between them.

The study of networks is known as graph theory, and, since its birth in 1736 by the Swiss mathematician Leonhard Euler, graph theory has solved several practical problems revealing interesting properties of many systems [3]. In the past decades, two important papers [4,5] established the mathematical bases of a new movement of interest and research in the study of complex networks, i.e., networks with irregular, complex, and dynamically evolving structure. The main focus was to provide tools for the analysis of irregular systems with thousands or millions of nodes.

The huge analysis of networks from different fields produced a series of unexpected and interesting results about their behaviors that led to the identification of a series of unifying principles and statistical properties. The effort was the definition of new concepts and measures to characterize the topology and the structure of real networks, to help in understanding their behavior and dynamic development. For example, it was shown how the robustness of the network systems to perturbations (such as failures and attacks) strongly depends on topology [6]; the functioning of the Internet network is studied from a topological point of view by Faloutsos et al. [7]; for the understanding of cancer cell growth mechanism, the analysis of the p53 gene and protein, and also the study of the whole network interacting with them, was proposed by Vogelstein et al. [8]. Consequently, a topological approach has become crucial in the last years, helping to characterize quantitatively certain local and global aspects of systems.

Based on the successful application of the Complex Network Theory to several fields (e.g., the Internet, neuroscience, computer science, biology, social science, medicine, etc.), in this paper, Water Distribution Networks (WDNs) are regarded as complex systems, modeled as graphs, and studied within the approach of the Complex Network Theory. In fact, WDNs can be considered as complex networks [9], as they are often constituted of thousands of elements, are strongly looped and show irregular shape, since they follow the layout of the city they serve. Water distribution networks have been successfully modeled as graphs to design an optimal sub-region layout [10], to study their global robustness [11–13], to evaluate their vulnerability to single pipe failures [14], to make a spectral analysis of the system [15,16], or to investigate possible benchmark values for the information entropy [17]. In general, strong correlations have been shown to exist among graph theory metrics and performance measures of model planar networks [18]; hence, the complex systems approach has been proposed as a framework to design and implement sustainable and hybrid water systems [19,20].

The complex and meshed structure of WDNs allows the system to recover from failures, exploiting the topological redundancy provided by closed loops, so that the flow could reach a given node through different paths. This redundant design approach gives the system an intrinsic capability of overcoming perturbations (e.g., local pipe failures or peaks of water demand), and, together with pipe diameters larger than those strictly necessary to fulfill the design pressure at the network nodes, to guarantee a power surplus to be dissipated [21,22]. In [21], Todini introduced a resilience index which accounts for the power surplus for an assigned network layout without any information about topology. More recently, Prasad and Park [23] proposed the concept of network resilience, which combines the effects of surplus power and reliable loops, but with a very simplified approach. Even if there is awareness that topology affects the performance and behavior of WDNs, this effect has not been quantified yet. Defining the topology of a WDN is a layout problem aimed at ensuring robustness and reliability [24] and represents one of the most difficult tasks in the design [25].

In fact, WDNs are critical infrastructures, whose reliable, robust, and efficient operation greatly affects national economic prosperity and people everyday life. It has been widely agreed that there are inseparable interdependencies between the robust structure of WDNs, their efficiency, as well as that of directly and indirectly related infrastructure operations. Therefore, the understanding of complex infrastructure systems needs a balance between holistic and reductionist methodologies [26]. This requires predictability of the complex network topology, behavior, and evolution dynamics over time, through a novel analysis framework.

In the last years, several new methodologies and metrics have been proposed in the scientific literature to better understand, describe, measure and optimize the topology of complex networks (a wide review is provided in [3,27]). Previous studies on real infrastructure networks indicated that they do not present small-world features (i.e., the presence of short geodesic distances between each nodes, meaning that it takes only a few steps to go from one node to another [4], and so the most part of the communication cross these hub nodes [5]). This is because, generally, infrastructure networks are planar networks with significant spatial limitation.

In this paper, the major and basic complex network metrics are applied to a large set of real and synthetic WDNs of various size. Then, the relationships between the obtained values of the metrics and the topology of WDNs are analyzed. The focus is identifying typical values for the topological metrics of WDNs to define a range of benchmark values and a general model of WDN structure, seeking possible relationships among these metrics and some indices of performance of WDNs. This topological framework could be a preliminary and alternative tool to the classical methods of analysis and design of WDNs, especially in the case of incomplete or lacking information about the system, since a calibrated hydraulic simulation model is not required. Furthermore, two well-known types of networks are considered, exploring the same number of nodes as the studied WDNs, namely the square grid and the random graph. For these networks, the topological metrics are calculated and used as a comparison to find possible relations and benchmark values. The results can help to define a mathematical model to describe the structure and the topology of WDNs. Finally, the clear relationships between some of the adopted topological metrics allows limiting the set of metrics needed to successfully describe the taxonomy of WDNs and so which of them should be further investigated and adopted as efficiently made for power grids by [28]. In this way, the paper can be seen as a contribution to the understanding, from a topological point of view, of WDNs, which could serve as a guidance for planning and monitoring practices.

2. Methods

From a topological and mathematical point of view, WDNs can be modeled as link-node planar (e.g., networks forming vertices wherever two edges cross) spatially organized weighted graphs $G = (V, E, w)$, where junctions, water sources and water demand points are represented by the set V of n nodes (hereinafter assumed as a measure of network size), while pipes and valves are represented by the set E of m edges, and w is a function that assigns to each edge a weight characterizing the physical characteristics of the pipe or of the valve [14]. In particular, WDNs belong to the class of networks strongly constrained by their geographical embedding [3], for which connections between distant nodes unlikely to be found, due to physical and economic constraints [29]. In particular, the long range connections in a spatial network are constrained by the Euclidean distance, having important consequences on the network statistical properties. In addition, the number of edges that can be connected to a single node is limited by the physical space to connect them (it is evident for urban streets, where only a small number of streets can cross in an intersection).

Generally, a complete representation of a network is provided by its adjacency matrix A which indicates which of the vertices are connected (adjacent): element $a_{ij} = 1$ indicates that there is a link between nodes i and j, $a_{ij} = 0$ otherwise. For an undirected network, the A matrix is symmetric since $a_{ij} = a_{ji}$. A weighted graph can be represented by its weighted adjacency matrix W where $w_{ij} > 0$ indicates the intensity of the link between nodes i and j, and $w_{ij} = 0$ if nodes i and j are disconnected. In the case of WDNs, the weight of the links could be hydraulic and/or geometric characteristics of the pipes (e.g., length, diameter, hydraulic resistance, flow, etc.) if available. From the adjacency matrix, the Laplacian matrix $L = D - A$ can be defined [30], where $D = diag(k_i)$ and $k_i = \sum_j a_{ij}$ is called the degree of the node i. In the case of a weighted graph, $k_i = \sum_j w_{ij}$. The matrices A and L described above represent two of the major and most frequently used graph matrices, the spectra of which, together with topological metrics defined and computed from them appear in many real case applications. In the following, the definitions of several topological metrics used in the paper are given. It is worth highlighting that most quantify the connectivity and the communication rate within a network. Hence, their meaning is diametrically opposed if network sectorization is discussed in terms of its effect on water flows or on potential contaminant transmission.

2.1. Link Density q

The link density q is the ratio between the total number m of network edges and the maximum number of edges $m^* = n(n-1)/2$ of a network with n nodes:

$$q = \frac{2m}{n(n-1)} \tag{1}$$

For most real networks, the link density value is low [27], since they are sparse, indicating that they are not fully connected.

2.2. Average Node Degree k

One of the simplest, and most important characteristics of a node is its degree k_i, defined as the total number of edges concurring in the node. The node degree counts the number of neighbors of node. The average value of k_i over all nodes n:

$$\overline{K} = \frac{2m}{n} \tag{2}$$

providing immediate information on the organization and structure of network, and its connectivity. The higher is the value of the average node degree, the better is the communication between the nodes.

2.3. Diameter D

The diameter D is defined as:

$$D = \max d_{ij} \tag{3}$$

where d_{ij} is defined as the shortest path from node i to node j, computed as the number of edges along the shortest path connecting them (when there is no path between a pairs of nodes, the distance is assumed infinite). The diameter D is defined as the maximum shortest distance (the maximum geodesic length) between any pair of vertices [31]. It expresses how cohesive a system is.

2.4. Average Path Length l

The average path length l is the average number of steps along the shortest paths for all possible pairs of nodes in the network, determining the average degree of separation between any pair of nodes:

$$l = \frac{2 \sum d_{ij}}{n(n-1)} \tag{4}$$

It measures the mean distance between two nodes, averaged over all pairs of nodes [4]. The geodesic length provides an optimal path way, since one would achieve a fast transfer and save system resources. The average path length gives information about the flow communication between any pairs of nodes.

2.5. Spectral Radius (or Spectral Index) λ_1^A

The spectral radius λ_1^A corresponds to the largest eigenvalue of the adjacency matrix A of a graph and it is related to the mean value of vertex degrees, taking into account not only immediate neighbors of vertices but also the neighbors of the neighbors [32]. Spectral radius plays an important role in abstract models for computer virus spreading through a network. In particular, the smaller the radius, the larger the network robustness against the spread of viruses is [33]. In this regard, Wang et al. [28] showed that the epidemic threshold (i.e., once exceeded, the infection survives and becomes an epidemic) in virus spreading is proportional to $1/\lambda_1^A$. This fact can be explained as the number of walks in a connected graph is proportional to λ_1^A. The greater is the number of walks of a network, the easier is the spread of the "moving substance" through it. Conversely, the higher is the spectral radius, the better is the communication within a network.

2.6. Spectral Gap $\Delta\lambda^A$

The spectral gap $\Delta\lambda^A$ is the difference between the first and second eigenvalue of the adjacency matrix A. Low values of this spectral metric indicate the presence of bottlenecks (articulation points or bridges) in the network [34], which hence can be easily split into sub-regions by removing few nodes or links [6].

2.7. Algebraic Connectivity λ_2^L

The algebraic connectivity λ_2^L corresponds to the second smallest eigenvalue of graph Laplacian matrix L [30], and quantifies the strength of network connections even if the graph is sparse ("how strong" are network connections). Its properties are extensively discussed in [35] with regard to its application to the analysis of graph robustness in terms of node and link failures, and proneness to clustering. Consequently, the larger the algebraic connectivity is, the more difficult it is to split the network into independent components.

2.8. Eigengap Δ^L

The eigengap $\Delta^L(s)$ corresponds to the difference between the $(s+1)^{th}$ eigenvalue and the sth eigenvalue of the Laplacian matrix:

$$\Delta^L(s) = \lambda_{s+1}^L - \lambda_s^L \tag{5}$$

where s is the number of clusters in which the network is intrinsically shaped. Choosing the proper number s of clusters is a general problem for all clustering algorithms, and, in the case of water distribution networks, it constitutes the arduous problem of water network partitioning [36]. A tool which is particularly designed for spectral clustering [35], but can also be applied successfully to other clustering algorithms, is the eigengap heuristic, which chooses the number of proper clusters c_{opt} such that all eigenvalues $\lambda_1, \ldots, \lambda_{c_{opt}}$ are small, but $\lambda_{c_{opt}+1}$ is relatively large. In other words, a simple indication of the proper number of clusters c_{opt}, from a topological point of view, is given by the first eigengap which results significantly larger than the previous ones. An explanation for this procedure, based on perturbation theory, is that, in the ideal case of c completely disconnected clusters, the 0 eigenvalue has multiplicity c_{opt}, and there is a gap to the $(c_{opt}+1)^{th}$ eigenvalue, that is $\lambda_{c_{opt}+1} > 0$ [37]. It is worth highlighting that the more pronounced is the cluster structure in the network, the better is the eigengap works.

3. Materials

To represent the topology of typical water systems, publicly available datasets of real and synthetic WDNs were used. These model networks are reported in the literature and their data-files are accessible on-line. Furthermore, to conduct the analysis and explore the structural properties of WDNs from a wide size range, we also compared the value of the above introduced topological metrics with that concerning square grids and random regular graphs. In Table 1, the name, the number of nodes n, the number of pipes m, the type (real or synthetic) and the data-file sources are reported for all networks used in the paper.

Table 1. Name of network, number of nodes n, number of pipes m, number of loops r, type and data-file sources for all WDNs.

Name	Number of Nodes n	Number of Links m	Number of Loops r	Type	Source
Two Loop	6	8	2	synthetic	[38]
Two Reservoirs	10	17	8	synthetic	[39]
New York tunnel	19	42	24	synthetic	[40]
Goyang	22	30	9	synthetic	[41]
Anytown	22	43	22	synthetic	[42]
Blacksburg	30	35	6	synthetic	[43]
Hanoi	31	34	4	synthetic	[44]
Bakryan	35	58	24	synthetic	[45]
Fossolo	36	58	23	synthetic	[46]
Richmond Skelton	68	99	4	synthetic	[47]
Pescara	41	44	32	synthetic	[46]
BWSN2008-1	126	168	43	real	[48]
Skiathos	175	189	15	real	[49]
Parete	184	282	101	real	[14]
Villaricca	196	249	54	real	[14]
Monteruscello	205	231	27	real	[50]
Modena	268	317	50	real	[46]
Celaya	333	477	145	real	[51]
Castellamare	365	439	75	real	GORI Spa
D-Town	399	443	45	real	[52]
Balerma Irrigation	443	454	12	real	[53]
Oreto	462	792	331	real	[54]
Richmond	865	949	85	real	[47]
Giugliano	994	1077	84	real	[16]
Matamoros	1283	1651	369	real	[10]
Wolf Cordera Ranch	1782	1985	204	real	[55]
San Luis Rio Colorado	1890	2681	792	real	[10]
Exeter	1891	3032	1142	synthetic	[56]
Exnet	1891	2465	575	synthetic	[56]
Denia	6276	6555	280	real	Aqualia
E-Town	11063	13896	2834	real	[57]
Alcala	11473	12454	982	real	Aqualia
BWSN2008-2	12523	14822	2300	real	[48]
Chihuahua	34868	40330	5463	real	[10]
SG1	9	12	4	synthetic	Matlab
SG2	100	180	81	synthetic	Matlab
SG3	1024	1984	961	synthetic	Matlab
SG4	10000	19800	9801	synthetic	Matlab
SG5	34969	69564	34596	synthetic	Matlab
RG1	10	15	6	synthetic	Matlab
RG2	100	150	51	synthetic	Matlab
RG3	1000	1500	501	synthetic	Matlab
RG4	10000	15000	5001	synthetic	Matlab
RG5	35000	52500	17501	synthetic	Matlab

Square grids: A lattice graph, mesh graph, or grid graph is a graph whose drawing, embedded in some Euclidean space R^n, forms a regular tiling [3]. This type of graph may more shortly be called just a lattice, mesh, or grid. A particular type of two-dimensional $n \times n$ lattice graph (indicated with $G_{n,n}$, and also known as square grid graph) is the graph whose vertices correspond to the points in the plane with integer coordinates, x-coordinates being in the range $1 \ldots n$, y-coordinates being in the range $1 \ldots n$, and two vertices are connected by an edge wherever the corresponding points are at distance 1 from each other (see Figure 1a). In other words, it is a unit distance graph for the described point set. A two-dimensional grid graph, also known as a square grid graph, is an $n \times n$ lattice graph $G_{n,n}$. In this paper, the two-dimensional square grid graph was considered, with a number of nodes equal to $n = 9, 100, 1024, 10000,$ and 34969 (respectively, named SG1, SG2, SG3, SG4, and SG5). In this way, a benchmark value was obtained for the entire size range of the studied networks. In fact, the largest considered square lattice has a number of nodes similar to the largest WDN considered in the paper (Chihuahua, for which $n = 34868$).

Random graph: The systematic study of random graphs was initiated by Erdos and Renyi [58]. The term random graph refers to the disordered arrangement of links between nodes. In particular, they considered graphs obtained by uniform sampling of all possible graphs with *n* vertices and *m* edges. In practice, random graphs are generated by connecting couples of randomly selected nodes (prohibiting multiple connections), until the number of edges equals *m* (see Figure 1b). It is clear that a given graph is only one realization of all the possible combinations of connections. A particular class of random graphs is the random *k*-regular graph, for which each node has the same number of neighbors (e.g., every node has the same degree *k*). A 3-regular graph is known as a cubic graph. Some important characteristics are: (a) a graph is regular if and only if it exists an eigenvector of the Adjacency matrix *A* whose eigenvalue is the constant degree *k* of the graph; and (b) a regular graph of degree *k* is connected if and only if the eigenvalue *k* has multiplicity one. Since *k*-regular graphs are subject to constraints, to generate them efficiently while ensuring an unbiased sampling one can resort either to the algorithm, implementing the most general configuration model [59], or to a refinement of such algorithm [60].

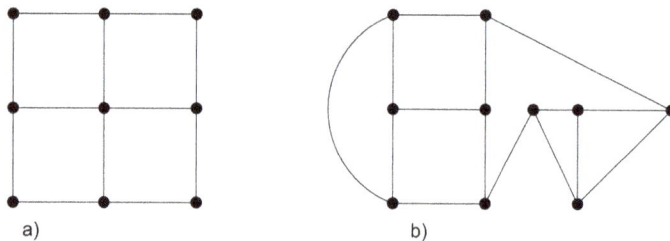

Figure 1. Examples of the synthetic networks used for comparison: (**a**) a small 3 × 3 square grid with 9 nodes and 12 edges; and (**b**) a small random 3-regular graph with 10 nodes and 15 edges.

In the present paper, to compare with the water networks explored, *k*-regular graphs with $k = 3$ and with a number of nodes equal to n = 10, 100, 1000, 10000, and 35000 were considered (respectively, named RG1, RG2, RG3, RG4, and RG5).

4. Results and Discussion

In recent years, several researchers studied the statistical and topological properties of several systems and infrastructures to provide novel solutions and understand what kind of network is needed to support and optimize the functioning of the systems themselves. In this regard, in this paper, the topological characteristics of WDNs are studied, based on some real-world and synthetic systems. In the following, the topological metrics defined in Section 2 are calculated for the networks described in Section 3, and then they are displayed as function of the number of nodes of each WDN. The results are reported in Table 2.

In Figure 2, the relationship between the link density *q* and the number of nodes *n* for the analyzed WDNs is plotted in log–log scale. It can be noticed that the two groups of WDNs (synthetic and real) follow the same trend with $q \sim n^{-1.04}$. Specifically, for increasing *n*, the link density tends to zero and closely follows a power-law with exponent −1, as reported in [27] for other real-world networks (linguistic systems, power grids, actor networks and biological systems). The link density shows a well defined scaling behavior, resulting inversely proportional to the network size *n*. It is worth highlighting that both square grids and random regular graphs show the same trend, meaning that water distribution networks are equally sparse in these two types of graph. In fact, link density is strongly related to the average node degree, which is not significantly different for WDNs, SG, and the considered RG. Such a similar behavior can be interpreted by means of Equation (1). In fact, introducing in Equation (1) the well-known topological relationship $m = n + r - 1$, linking the number of pipes m, nodes n, and loops r of a network, it results $q = 2/n + 2r/n(n-1)$. As in WDNs it is $r << (n-1)$, the second term is always smaller than the first, so that $q = 2/n$. SG and RG are instead more looped than most WDNs. In fact, in Figure 1, it is clear that the link density is slightly higher for SG and RG, which is due to the presence of a greater number of loops (according to the above relationship), as shown in Table 1. In this regard, since the number of loops can be regarded as a robustness metric, clearly the link density can be used as a surrogate metric for the robustness of WDNs. In fact, it takes into account the number of nodes, pipes and, implicitly, the number of loops.

Table 2. Topological metric values calculated for all case studies: link density q, average node degree $\overline{K}$, graph diameter D, average path length l, spectral gap $\Delta\lambda^A$, algebraic connectivity λ_2^L, inverse spectral radius $1/\lambda_1^A$, optimal cluster number c_{opt}.

Name	q	$\overline{K}$	D	l	$\Delta\lambda^A$	λ_2^L	$1/\lambda_1^A$	c_{opt}
Two Loop	0.5333	2.67	4	1.90	1.2213	0.68862	0.404	2
Two Reservoirs	0.3778	3.40	6	2.59	0.8799	0.37909	0.303	2
New York tunnel	0.2456	4.42	9	4.21	0.5560	0.11799	0.400	3
Goyang	0.1299	2.73	9	3.75	0.2595	0.09969	0.331	3
Anytown	0.1861	3.91	7	2.94	0.4581	0.28044	0.218	2
Blacksburg	0.0805	2.33	9	4.37	0.3077	0.08998	0.372	2
Hanoi	0.0731	2.19	13	5.31	0.2739	0.06116	0.412	4
Bakryan	0.0975	3.31	12	4.30	0.4793	0.07860	0.299	3
Fossolo	0.0921	3.22	8	3.67	0.3516	0.21888	0.307	2
Richmond Skelton	0.0537	2.15	24	9.24	0.0266	0.01091	0.411	3
Pescara	0.0435	2.91	20	8.69	0.3024	0.00891	0.306	4
BWSN2008-1	0.0213	2.67	25	10.15	0.1004	0.00750	0.330	4
Skiathos	0.0124	2.16	27	11.52	0.0461	0.00835	0.374	3
Parete	0.0171	3.10	20	8.80	0.1714	0.02117	0.303	4
Villaricca	0.0130	2.54	32	11.29	0.1194	0.00665	0.334	4
Monteruscello	0.0110	2.25	47	20.24	0.0481	0.00152	0.352	5
Modena	0.0089	2.37	38	14.04	0.1385	0.00908	0.334	6
Celaya	0.0086	2.86	32	11.81	0.1915	0.01336	0.281	6
Castellamare	0.0066	2.41	37	13.62	0.1640	0.00627	0.311	6
D-Town	0.0056	2.22	66	26.32	0.0703	0.00065	0.350	5
Balerma Irrigation	0.0046	2.05	60	23.89	0.0845	0.00069	0.370	6
Oreto	0.0074	3.43	27	11.98	0.2016	0.00492	0.252	4
Richmond	0.0025	2.19	135	51.44	0.0727	0.00014	0.345	8
Giugliano	0.0022	2.17	51	21.22	0.1354	0.00243	0.327	9
Matamoros	0.0020	2.57	80	27.76	0.1439	0.00100	0.291	8
Wolf Cordera Ranch	0.0013	2.23	69	25.94	0.0612	0.00053	0.326	8
San Luis Rio Colorado	0.0015	2.84	76	28.86	0.0063	0.00089	0.268	7
Exeter	0.0017	3.21	54	20.61	0.6121	0.01021	0.257	10
Exnet	0.0014	2.61	59	21.31	0.1190	0.00102	0.257	10
Denia	0.0003	2.09	186	70.48	0.0797	0.00004	0.328	17
E-Town	0.0002	2.51	289	71.13	0.0570	0.00003	0.281	13
Alcala	0.0002	2.17	163	64.88	0.0957	0.00009	0.295	13
BWSN2008-2	0.0002	2.37	297	93.30	0.0147	0.00002	0.308	14
Chihuahua	0.0001	2.31	368	186.05	0.0175	0.00001	0.282	18
SG1	0.3333	2.67	4	2.00	1.4142	1.00000	0.354	-
SG2	0.0364	3.60	18	6.67	0.2365	0.09790	0.261	-
SG3	0.0037	3.88	62	21.33	0.0271	0.00963	0.251	-
SG4	0.0004	3.96	198	66.67	0.0029	0.00099	0.250	-
SG5	0.0001	3.98	398	124.67	0.0008	0.00028	0.250	-
RG1	0.3333	3.00	3	1.80	1.3820	1.38200	0.333	-
RG2	0.0303	3.00	8	4.73	0.2614	0.26142	0.333	-
RG3	0.0030	3.00	13	8.04	0.1805	0.18055	0.333	-
RG4	0.0003	3.00	17	11.37	0.1737	0.17368	0.333	-
RG5	0.0001	3.00	19	13.01	0.1717	0.17170	0.333	-

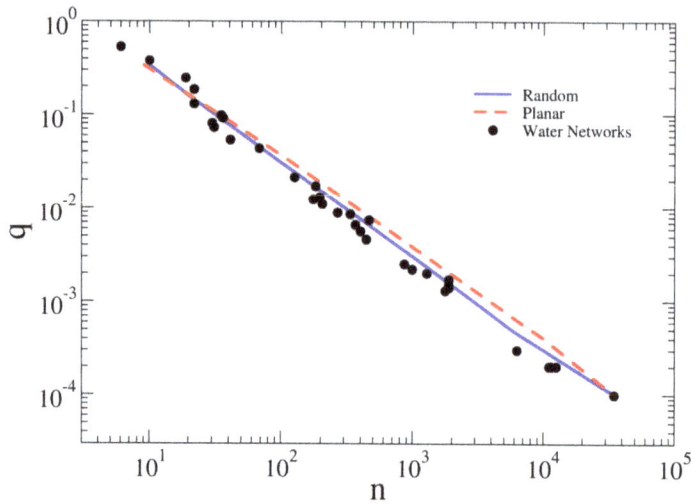

Figure 2. Relationship between the link density q and the network size n. The data points follow a trend that can be fitted with a power law decrease $q \propto n^{-1.4}$. Both water networks, random regular graphs and planar square lattice show a similar $q \sim \frac{1}{n}$ scaling expected for sparse networks. The blue continuous line and the broken red line refer to random graph and square grids, respectively. The black dots represent the studied WDNs. For numerical values, please see Table 2.

In Figure 3, the average node degree $\overline{K}$—the coarsest characteristic of node interconnections [27]—is plotted in semi-log scale as a function of the size n. We observe that the average node degree is nearly invariant with respect to the network size n—perhaps with a slightly decreasing trend—confirming that water distribution networks are sparsely connected. Such a behavior is expected for real networks with economic and geographic constraints [16,29]. It is possible to identify a range of values from $\overline{K} \sim 2$ to $\overline{K} \sim 4.5$ that are the typical small values of WDNs. In fact, the lower bound ($\overline{K} = 2$) corresponds to simple line graphs, which have the lowest topological robustness, as the failure of a single pipe leads to a complete network disconnection. An important aspect is that the nearly invariant trend is also observed for the square grids, for which the average node degree seems to tend, as the number of nodes increases, to an asymptotic value of $\overline{K} = 4$. For the random graph, the average node degree is obviously constant and equal to $\overline{K} = 3$ for all networks (cubic graphs were chosen for the comparison). The small variations of the average node degree for the two groups of WDNs (synthetic and real) is another aspect of the n^{-1} trend observed for link density in Figure 2; in fact, $q = 2m/[n(n-1)] = \overline{K}/(n-1)$. Another important aspect to highlight is that the node degree distribution is almost homogeneous [61], i.e., almost all nodes have the same degree.

Hence, water distribution networks are not characterized by the presence of hubs—nodes with very high degree—that happens in the case of scale free networks. This leads to the immediate consequence that generally WDNs are almost equally robust against both intentional and accidental pipe breaks. In this case, by taking into account the relationship $m = n + r - 1$, it is possible to better interpret the typical values shown by WDNs. It results $\overline{K} = 2 + 2(r-1)/n$, which implies that $\overline{K} = 2$ is the lowest possible value for WDNs (i.e., no loops), and that values above 2 are related to the number of loops and thus to the ratio r/n. In this respect, the average node degree can be used as a surrogate measure of robustness, quantifying how a WDN is far from a line shaped network (for which $\overline{K} = 2$).

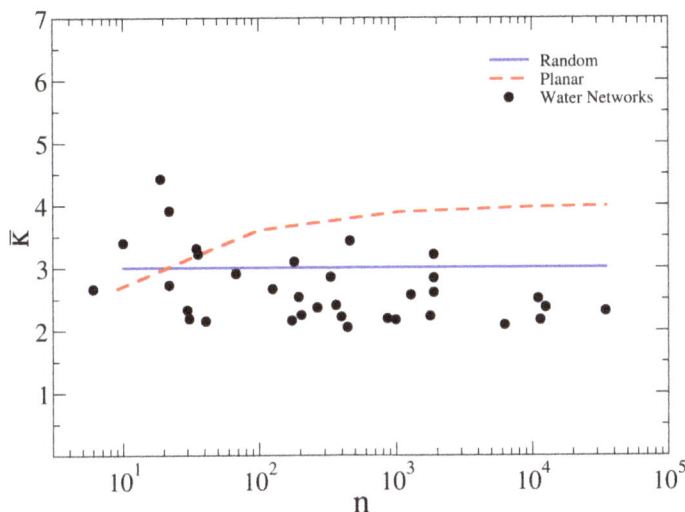

Figure 3. Relationship between the average node degree $\overline{K}$ and the network size n compared to the case of random regular networks and planar square grids. Notice that, due to boundary effects, the square grids approach their theoretical value $\overline{K} \sim 4$ only for large network sizes n. The blue continuous line and the broken red line refer to random graph and square grids, respectively. The black dots represent the studied WDNs. For numerical values, please see Table 2.

In Figure 4, the diameter D is plotted in log–log scale as a function of the number of nodes n. It is clear that the graph diameter, for both synthetic and real WDNs, increases as the network size increases, following a power law, with $D \propto n^{0.51}$. Thus, differently from communication systems, WDNs do not represent the peculiar features of Small-World networks, for which—similar to in random networks—$D \propto log(n)$. Such a feature leads to robustness in communications, since the average shortest path between two nodes increases very slowly with the network size. The larger diameter graph for the WDNs is due to economic and physical/geographic constraints that generally do not allow the presence of long-range links, except in rare cases [29]. In fact, it is well known that even a small fraction of links between distant nodes in a network can significantly shorten the path lengths [4]. It is clear that, also from a communication point of view, square grids and WDNs show a similar structure. In fact, in the SG, the nodes are always linked to the adjacent neighborhoods. It means that, from a global communication point of view, the possible presence of few long-range links in WDNs is not enough to reduce D, which also corresponds to the maximum of the shortest paths between nodes. Conversely, the random regular graphs, with the same number of nodes and with an average node degree close to WDNs, show a significantly lower diameter. In particular, they can be well fitted by a $D \propto log(n)$, as expected in general for random graphs [58].

In Figure 5, the average path length l is plotted in log–log scale as a function of the number of nodes n. The trend clearly resembles that of the diameter, as shown in Figure 3. In particular, l increases as the network size increases following a power law, with $l \propto n^{0.48}$, for the same reasons described above. In this case, the square grids follow a trend very close to WDNs, while the random graphs confirm their better flow communication, showing a logarithmic increase of the average path length. It can be observed that the $n^{1/2}$ scaling of D and l in WDNs reflects their embedding in a 2D-spatial environment. Very differently, the randomness of the cubic graphs provides the possibility to also have long-range connections that drastically reduce the length of the shortest paths between each pair of nodes. In this case, the global communication of the WDNs is strongly influenced by the fact that most nodes are linked to the adjacent neighborhoods, which makes them very similar to the SG, for which all the nodes are always and only linked to the adjacent neighborhoods.

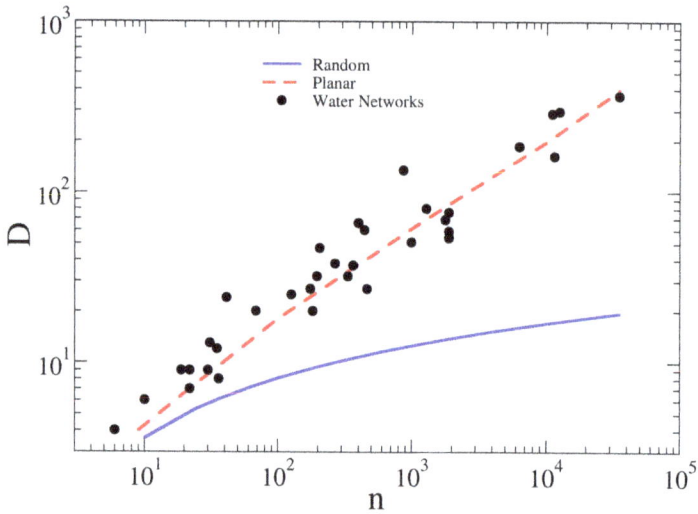

Figure 4. Plot of the diameter D versus the network size n for synthetic and real water networks, planar square grids and random graphs. Notice that, similar to planar square grids, water networks also show the $D \sim n^{1/2}$ scaling that is expected for planar spatial networks. The blue continuous line and the broken red line refer to random graph and square grids, respectively. The black dots represent the studied WDNs. For numerical values, please see Table 2.

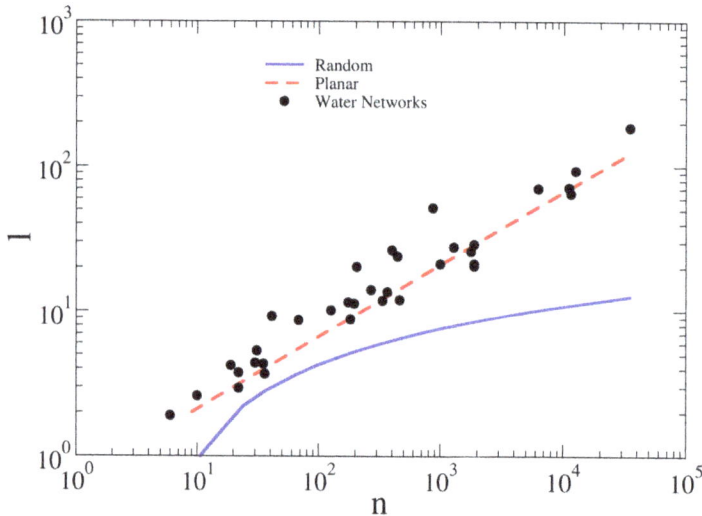

Figure 5. Plot of the average path length l versus the network size n for synthetic and real water networks, square grids and random graphs. Notice that, similar to planar square grids, water networks also show the $l \sim n^{1/2}$ scaling that is expected for planar spatial networks. The blue continuous line and the broken red line refer to random graph and square grids, respectively. The black dots represent the studied WDNs. For numerical values, please see Table 2.

In Figure 6, the inverse of the spectral index λ_1^A is plotted in semi-log scale as a function of the water distribution network size n. It is possible to identify a typical WDN range with $(\lambda_1^A)^{-1} \approx 0.3 \pm 0.1$, and so for λ_1^A, which is related to the general relationship according to which $\overline{K} \leq \lambda_1^A \leq \max(k)$

[33]. For this reason, the inverse spectral radius of cubic random graphs assumes the constant value $(\lambda_1^A)^{-1} = 0.33$, while for square grids it tends asymptotically to $(\lambda_1^A)^{-1} = 0.25$ (for these networks the maximum node degree is equal to 4 and the larger the number of the network nodes, the fewer of them are boundary nodes, which have a degree smaller than 4). Finally, for the water distribution networks, λ_1^A is nearly invariant with the size of the network. It means that, even if the size of the water network increases, the topology does not vary. Since the spectral radius is linked to the "velocity of spread" of a substance in the presence of percolation dynamics, it means that for WDNs it is possible to define a characteristic value of the intrinsic capacity of being crossed by a substance, water flow or contaminant. In this regard, the nearly invariant value of the spectral index suggests that, in general, the probability of the spreading of a contaminant from a point to another in a WDN does not depend on its dimension, as, apart from the hydraulic characteristics of the system, it is strongly related to the number of connections of each node. The value of the spectral index for WDNs falls between that of RG and SG as the number of nodes increases. It can be observed that the SG are the most vulnerable in terms of substance spreading, since most of the nodes have a degree equal to 4, while the RG, having a degree equal to 3 for all nodes, show the lowest value of $(\lambda_1^A)^{-1}$. For WDNs, even if the average node degree is usually lower than RG, the small fraction of nodes with a degree higher than 3 gives WDNs a higher capability of being crossed by a substance. It can also be observed that the relation $\overline{K} \leq \lambda_1^A \leq \max(k)$ indicates that the spectral radius simultaneously quantifies global and local connectivity of the network. In fact, the lowest bound is related to the number of loops, while the highest is related to the presence of hubs.

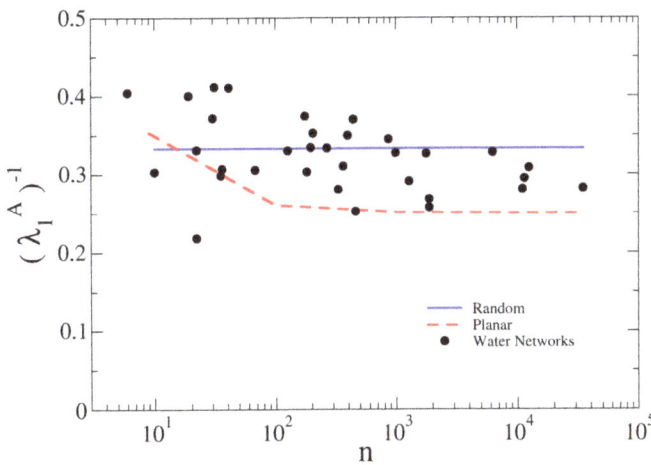

Figure 6. Relationship between the inverse of the spectral index λ_1^A and the network size n for water networks, random cubic networks and planar square grids. The blue continuous line and the broken red line refer to random graph and square grids, respectively. The black dots represent the studied WDNs. For numerical values, please see Table 2.

In Figure 7, the behavior of the spectral gap $\Delta\lambda^A$ is plotted in log–log scale as a function of the network size n. A decreasing trend of the spectral metric with increasing size is visible for all the studied networks, which could be fitted with a power scaling law $\Delta\lambda^A \sim n^{-0.36}$. While, for square grids and random graphs, the trends are quite clear—respectively, a power law decay and an exponential decay to a plateau—for WDNs, the trend with network size is less clearly defined. However, the dots representing the studied WDNs fall between the trend lines of square grids and random graphs. As a small $\Delta\lambda^A$ indicates high probability of having articulation points or bridges (the failure of which can cause the disconnection of the network in more sub-regions), this result points out that large WDNs tend to be increasingly less robust against disconnection. It is also clear that random graphs show a higher robustness than the square grids, since they show a higher value of the spectral gap $\Delta\lambda^A$. This is due first to the fact that the chosen cubic graphs have constant average node degree $\overline{K} = 3$ which guarantees that all nodes are linked to other three nodes. Furthermore, their randomness ensures the presence of long-range links that lead to a more cohesive and compact structure. Clearly, this metric does not depend on the number of connections, but rather on how nodes are connected to each other. In fact, the SG have highest $\overline{K}$, but at the same time they show the smallest spectral gap. In this respect, the possible presence of few long-range links gives some randomness to WDNs, thus resulting more similar to RG. This behavior resembles the concept of Pseudorandom graph (e.g., Torres et al. [18]), in which random connections are allowed only between neighboring nodes.

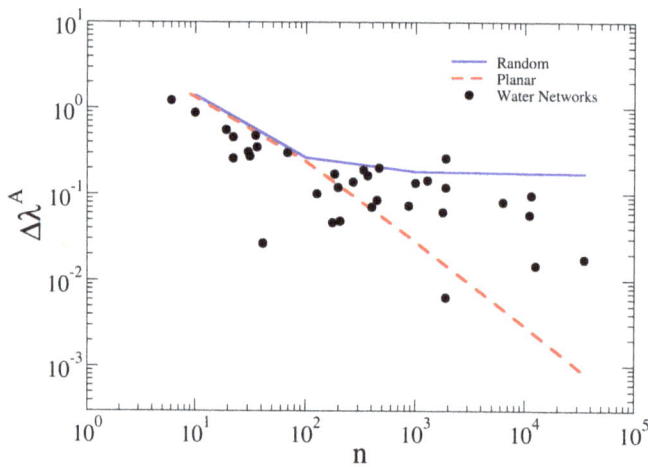

Figure 7. Relationship between the spectral gap $\Delta\lambda^A$ and the network size n. For the water networks, the fit of the data points is a power law $\Delta\lambda^A \sim n^{-0.36}$; for the random networks, thick line is fitting to an exponential approach $\sim e^{-n/35}$ to a plateau value of $\Delta\lambda^A$; and, for the square grids, to a power law $\Delta\lambda^A \sim n^{-0.93}$. The blue continuous line and the broken red line refer to random graph and square grids, respectively. The black dots represent the studied WDNs. For numerical values, please see Table 2.

In Figure 8, the behavior of the algebraic connectivity λ_2^L is plotted in log–log scale as a function of the network size n. A clear decrease of the connectivity metric with increasing size can be seen for all the networks, confirming the results of the spectral gap, but with clearer trends. Specifically, the water distribution networks and the square grids show a power law behavior, respectively, $\lambda_2^L \propto n^{-1.26}$ and $\lambda_2^L \propto n^{-0.99}$, thus both curves tend to zero for increasing system size. This implies that the robustness of these networks, and in particular of WDNs, decreases as the number of nodes increase, with a high probability to have bottlenecks that can be easily broken with small effort (low value of λ_2^L). The algebraic connectivity of WDNs results always smaller than for the square grids, implying in general lower robustness, owing to the smaller number and less regularity of the connections between nodes. On the other hand, the randomness of the k-regular graphs provides a higher robustness to the system: in fact, the algebraic connectivity shows an exponential trend $\sim e^{-n/35}$, approaching a plateau value of $\lambda_2^L = 0.17$; hence, fragility does not increase with system size but stabilizes at a constant value. Hence, the algebraic connectivity indicates that large WDNs can be easily subdivided (i.e., create clusters with high density intra-clusters and low density infra-clusters), meaning that, as suggested by

Wang et al. [28] for power grids, WDNs show a nested layout. In other words, it is easy to identify regions with different levels of density that can be isolated from the rest of the network. Clustering seems to constitute an intrinsic topological property of WDNs.

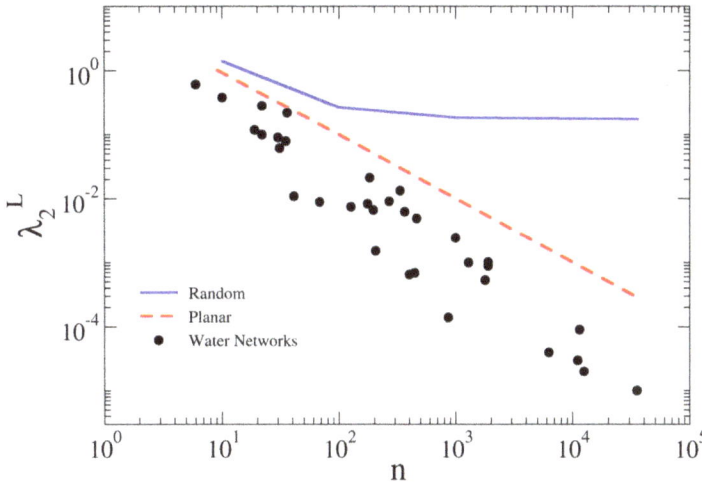

Figure 8. Relationship between the algebraic connectivity λ_2^L and the network size n. For the water networks, the fit of the data points is a power law $\lambda_2^L \sim n^{-1.26}$; for the random networks, thick line is fitting to an exponential approach $\sim e^{-n/35}$ to a plateau value of λ_2^L; and, for the square grids, to a power law $\lambda_2^L \sim n^{-0.99}$. The blue continuous line and the broken red line refer to random graph and square grids, respectively. The black dots represent the studied WDNs. For numerical values, please see Table 2.

Given a graph $G = (V, E)$, a community (or cluster, or cohesive subgroup) is a sub-graph $G = (V', E')$, whose nodes are tightly connected, i.e., cohesive. The community structure of the complex network systems constitutes a powerful tool for better understanding the functioning of the network itself, as well as for identifying a hierarchy of connections within a complex architecture. Different metrics can lead to different communities; however, for physical networks, one of the most natural methods to partition its graph in reasonable communities is spectral partitioning [62], since it allows the separation into subgraphs minimizing the number of links between such subgraphs. Spectral partitioning uses the eigenvectors of the Laplacian matrix of a graph to determine the subgraphs corresponding to separate communities. To optimize the number of such communities, it is customary to look at the *eigengap*, i.e., at the maximum jump in the spectrum of the Laplacian matrix. According to such a criterion, only the eigenvectors whose eigenvalue is smaller than the eigengap are used to partition the network; in this regard, it has been shown that the eigengap can constitute a valid and useful tool to solve the problem of establishing a preliminary number of districts for the water network partitioning [16], according to only topological criteria, especially when no other information is available. In Figure 9, the first largest eigengap $\Delta^L(s)$ of the Laplacian matrix is used to calculate the optimal number of cluster c_{opt} as a function of the network size n. It looks clear how the number of clusters c_{opt} in which the water distribution networks are divided according to Δ^L increases with the system size. In particular, it follows approximatively a power law $c_{opt} \propto n^{0.28}$. Hence, the number of districts grows sub-linearly with the network size, indicating that the optimal number of districts does not increase significantly with system size. It is worth noticing that, since c_{opt} grows sub-linearly, the number of partitions grows more slowly than the number of elements of the network. Hence, for large water networks, the size of optimal districts grows, compared to small water networks. It is worth noting that this result holds for network partitioning only from the topological point of view, while also the aims of the sectorization of the network should be considered in the choice of the optimal dimension of districts. In fact, it is known that the larger districts are, the more difficult is the identification of bursts and leakages from night flow data, although smaller districts imply higher costs for valves, flow meters and maintenance. Conversely, in terms of network safety against accidental or intentional water contamination, the larger is the district, the larger is the number of users potentially exposed to injected pollutants.

However, the obtained result points out that, in large WDNs, the size of the districts should necessarily increase. Otherwise, the benefits of an easier management guaranteed by smaller districts would be partly nullified by the increased vulnerability (or lower robustness) of the network, caused by the excessive fragmentation.

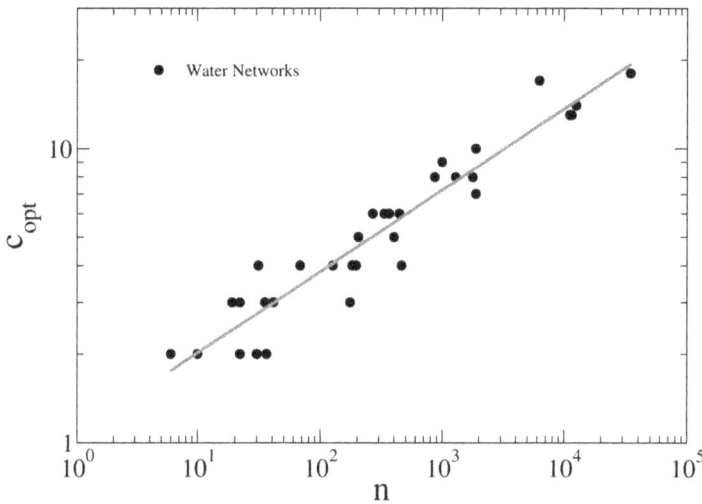

Figure 9. Relationship between the optimal number of clusters c_{opt} and the network size n. The thick brown line is a power law fit for the analyzed water networks, yielding $c_{opt} \sim n^{0.28}$, providing an indication for the optimal number of District Metered Areas (DMAs) of a water network of given size. The black dots represent the studied WDNs. For numerical values, please see Table 2.

5. Conclusions

The topological analysis of several real and synthetic water distribution networks shows that such networks tend to be sparse, being characterized by small values of the average degree $\overline{K}$ and of the link density $q \sim n^{-1}$. The nearly homogeneous value of the node degree marks a distance between WDNs and scale-free networks. The studied networks show many characteristics of planar lattices, since both the diameter D and the average path length l scale as $n^{1/2}$. Such power law trends confirm that WDNs, similar to power grids or street networks, cannot be modeled as small world systems. Summarizing, all the evaluated topological connectivity metrics indicate that the graphs of WDNs are far from being totally random, as often claimed in theoretical studies, and rather resemble regular square grids.

The analysis of the spectral metrics, however, points out that WDNs present some randomness which can have a positive effect on their topological robustness, despite the geographical constraints, which make them close to planar graphs. In fact, for large network size, both the spectral index λ_1^A and the spectral gap $\Delta\lambda^A$ fall in between the values of random graphs and square grids. On the one side, the spectral index λ_1^A, similar to planar lattices (for which $\lambda_1^A \sim \overline{K}$), results nearly constant with network size, indicating that WDNs are topologically protected from the spread of a contaminant, apart from the hydraulic characteristics of the pipes. On the other side, the decreasing trend of the spectral gap $\Delta\lambda^A$ with the network size indicates the presence of bottlenecks and articulation points, but not leading to the complete disintegration of the network, as for regular square grids. In fact, while the strongly decreasing trend of the algebraic connectivity λ_2^L with the network size indicates that the "energy" required to break the network into independent sub-regions becomes lower, the optimal number of clusters c_{opt}, identified by means of the eigengap $\Delta^L(s)$, grows less than linearly with the network size. Such a sub-linear growth hints that, from a connectivity point of view, the larger the WDN, the larger is the optimal size of DMAs.

The topological analysis of an extensive number of real and synthetic water distribution networks indicates that it is possible to identify a limited set of metrics that completely characterize the topological structure of WDNs. In particular, the average node degree $\overline{K}$ strongly influences the values of the spectral index λ_1^A and of the link density q. Regarding the communication metrics, it is evident that the graph diameter D and the average path length l provide nearly the same information about the topology of a WDN. It seems preferable to use l, since it expresses a mean value over all paths, and because it is more sensitive than D to the addition or removal of an edge. The comparison between the two spectral robustness metrics, the spectral gap $\Delta\lambda^A$ and the algebraic connectivity λ_2^L, suggests that, in the case of WDNs, the latter is more significant, both because it is strongly related to the strength needed to split the network into sub-regions, and because it shows a clearer trend with the network size. Finally, from a topological connectivity point of view, the eigengap $\Delta^L(s)$ provides a quick and good estimate of the optimal number of districts for the partitioning of a WDN.

According to the results presented in this study, the topological structure of WDN is very far from the "totally" random networks often used in theoretical studies. Similar to the concept of Nested-Smallworld, introduced for power-grids [28], and to the concept of Pseudorandom graph, introduced by Torres et al. [18], WDNs could be classified as a novel structure defined as Nested-Pseudorandom graph, because they show simultaneously nested and pseudorandom characteristics. In particular, such a model is the result of connecting several pseudorandom sub-networks through few long-range links, also in accordance with the above mentioned intrinsic clustering property of WDNs.

The typical values of the topological metrics calculated in this paper could be used to generate graphs of synthetic water distribution networks, which retain the topological characteristics of real WDNs, e.g., through a graph generating software. This would allow having many test cases for modeling purposes (it is not always easy to have the data and the graph of real WDNs), with realistic topologies and with network size. In this respect, the automatic generation would be a useful tool, as the currently used synthetic WDNs generally have small dimensions.

Acknowledgments: AS thanks CNR-PNR National Project "Crisis-Lab" for support. The contents of the paper do not necessarily reflect the position or the policy of funding parties. This research is part of the Ph.D. project "Water distribution network management optimization through Complex Network theory" within the Doctoral Course "A.D.I." granted by Universita' degli Studi della Campania "L. Vanvitelli".

Author Contributions: All authors contributed equally to the paper.

Conflicts of Interest: The authors declare no conflict of interest. The founding sponsors had no role in the design of the study; in the collection, analyses, or interpretation of data; in the writing of the manuscript, and in the decision to publish the results.

Abbreviations

The following abbreviations are used in this manuscript:

WDN	Water Distribution Network
CWDN	Complex Water Distribution Network
DMAs	District Metered Areas

References

1. Dorogovtsev, S.N.; Mendes, J.F.F. *Evolution of Networks: From Biological Nets to the Internet and WWW (Physics)*; Oxford University Press, Inc.: New York, NY, USA, 2003.

2. D'Agostino, G.; Scala, A. (Eds.) *Networks of Networks: The Last Frontier of Complexity*; Understanding Complex Systems; Springer: Berlin, Germany, 2014.
3. Boccaletti, S.; Latora, V.; Moreno, Y.; Chavez, M.; Hwang, D.U. Complex networks: Structure and dynamics. *Phys. Rep.* **2006**, *424*, 175–308.
4. Watts, D.J.; Strogatz, S.H. Collective dynamics of 'small-world' networks. *Nature* **1998**, *393*, 440–442.
5. Barabasi, A.; Albert, R. Emergence of Scaling in Random Networks. *Science* **1999**, *286*, 509–512.
6. Estrada, E. Network robustness to targeted attacks. The interplay of expansibility and degree distribution. *Eur. Phys. J. B Condens. Matter Complex Syst.* **2006**, *52*, 563–574.
7. Faloutsos, M.; Faloutsos, P.; Faloutsos, C. On power-law relationships of the Internet topology. *Comput. Commun. Rev.* **1999**, *29*, 251–262.
8. Vogelstein, B.; Lane, D.; Levine, A.J. Surfing the p53 network. *Nature* **2000**, *408*, 307–310.
9. Mays, L.W. *Water Distribution Systems Handbook*; McGraw-Hill: New York, NY, USA, 2000.
10. Tzatchkov, V.G.; Alcocer-Yamanaka, V.H.; Ortiz, V.B. Graph Theory Based Algorithms for Water Distribution Network Sectorization Projects. In Proceedings of the Eighth Annual Water Distribution Systems Analysis Symposium (WDSA), Cincinnati, OH, USA, 27–30 August 2006.
11. Yazdani, A.; Jeffrey, P. Robustness and Vulnerability Analysis of Water Distribution Networks Using Graph Theoretic and Complex Network Principles. In Proceedings of the 12th Annual Conference on Water Distribution Systems Analysis (WDSA), Tucson, AZ, USA, 12–15 September 2010.
12. Yazdani, A.; Jeffrey, P. Complex network analysis of water distribution systems. *Chaos Interdiscip. J. Nonlinear Sci.* **2011**, *21*, 016111,
13. Di Nardo, A.; Di Natale, M.; Giudicianni, C.; Musmarra, D.; Santonastaso, G.F.; Simone, A. Water Distribution System Clustering and Partitioning Based on Social Network Algorithms. *Procedia Eng.* **2015**, *119*, 196–205.
14. Di Nardo, A.; Di Natale, M.; Giudicianni, C.; Musmarra, D.; Rodriguez Varela, J.; Santonastaso, G.; Simone, A.; Tzatchkov, V. Redundancy Features of Water Distribution Systems. *Procedia Eng.* **2017**, *186*, 412–419.
15. Herrera, M.F. Improving Water Network Management by Efficient Division into Supply Clusters. Ph.D. Thesis, Universitat Politécnica de Valencia, Valéncia, Spain, 2011.
16. Di Nardo, A.; Di Natale, M.; Giudicianni, C.; Greco, R.; Santonastaso, G.F. Water Supply Network Partitioning Based On Weighted Spectral Clustering. In *Complex Networks & Their Applications V: Proceedings of the 5th International Workshop on Complex Networks and their Applications (COMPLEX NETWORKS 2016)*; Cherifi, H., Gaito, S., Quattrociocchi, W., Sala, A., Eds.; Springer: Cham, The Netherland, 2017; pp. 797–807.
17. Santonastaso, G.F.; Di Nardo, A.; Di Natale, M.; Giudicianni, C.; Greco, R. Scaling-Laws of Flow Entropy with Topological Metrics of Water Distribution Networks. *Entropy* **2018**, *20*, 95.
18. Torres, J.M.; Duenas-Osorio, L.; Li, Q.; Yazdani, A. Exploring Topological Effects on Water Distribution System Performance Using Graph Theory and Statistical Models. *J. Water Resour. Plan. Manag.* **2017**, *143*, 04016068.
19. Facchini, A.; Scala, A.; Lattanzi, N.; Caldarelli, G.; Liberatore, G.; Maso, L.D.; Nardo, A.D. Complexity Science for Sustainable Smart Water Grids. In *Italian Workshop on Artificial Life and Evolutionary Computation*; Springer: Cham, The Netherland, 2017; Volume 708.
20. Zodrow, K.R.; Li, Q.; Buono, R.M.; Chen, W.; Daigger, G.; Duenas-Osorio, L.; Elimelech, M.; Huang, X.; Jiang, G.; Kim, J.H.; et al. Advanced Materials, Technologies, and Complex Systems Analyses: Emerging Opportunities to Enhance Urban Water Security. *Environ. Sci. Technol.* **2017**, *51*, 10274–10281,
21. Todini, E. Looped water distribution networks design using a resilience index based heuristic approach. *Urban Water* **2000**, *2*, 115–122.
22. Di Nardo, A.; Di Natale, M.; Giudicianni, C.; Santonastaso, G.; Savic, D. Simplified Approach to Water Distribution System Management via Identification of a Primary Network. *J. Water Resour. Plan. Manag.* **2018**, *144*, 04017089.
23. Prasad, T.D.; Park, N.S. Multiobjective Genetic Algorithms for Design of Water Distribution Networks. *J. Water Resour. Plan. Manag.* **2004**, *130*, 73–82.
24. Wagner, J.M.; Shamir, U.; Marks, D.H. Water Distribution Reliability: Analytical Methods. *J. Water Resour. Plan. Manag.* **1988**, *114*, 253–275.
25. Goulter, I.C.; Bouchart, F. ReliabilityConstrained Pipe Network Model. *J. Hydraul. Eng.* **1990**, *116*, 211–229.
26. Rouse, W.B. Complex engineered, organizational and natural systems. *Syst. Eng.* **2007**, *10*, 260–271.

27. Jamakovic, A.; Uhlig, S. On the relationships between topological measures in real-world networks. *Netw. Heterog. Media* **2008**, *3*, 345–359.

28. Wang, Z.; Scaglione, A.; Thomas, R. Generating Statistically Correct Random Topologies for Testing Smart Grid Communication and Control Networks. *IEEE Trans. Smart Grid* **2010**, *1*, 28–39.

29. Amaral, L.A.N.; Scala, A.; Barthélémy, M.; Stanley, H.E. Classes of small-world networks. *Proc. Natl. Acad. Sci. USA* **2000**, *97*, 11149–11152.

30. Fiedler, M. Algebraic Connectivity of Graphs. *Czech. Math. J.* **1973**, *23*, 298–305.

31. Watts, D.J. *Small Worlds: The Dynamics of Networks Between Order and Randomness*; Princeton University Press: Princeton, NJ, USA, 1999; p. xv, 262p.

32. Bonacich, P. Power and Centrality: A Family of Measures. *Am. J. Sociol.* **1987**, *92*, 1170–1182.

33. Arsic, B.; Cvetkovic, D.; Simic, S.K.; Skaric, M. Graph spectral techniques in computer sciences. *Appl. Anal. Discret. Math.* **2012**, *6*, 1–30.

34. Donetti, L.; Neri, F.; Munoz, M.A. Optimal network topologies: Expanders, cages, Ramanujan graphs, entangled networks and all that. *J. Stat. Mech. Theory Exp.* **2006**, *2006*, P08007.

35. Von Luxburg, U. A Tutorial on Spectral Clustering. *Stat. Comput.* **2007**, *17*, 395–416.

36. Perelman, L.S.; Michael, A.; Ami, P.; Mudasser, I.; Whittle, A.J. Automated sub-zoning of water distribution systems. *Environ. Model. Softw.* **2015**, *65*, 1–14.

37. Chung, F.R.K. *Spectral Graph Theory*; American Mathematical Society: Providence, RI, USA, 1997.

38. Alperovits, E.; Shamir, U. Design of optimal water distribution systems. *Water Resour. Res.* **1977**, *13*, 885–900.

39. Gessler, J. Pipe network optimization by enumeration; In *Proceedings of Computer Applications for Water Resources, New York*; ASCE: Reston, VA, USA, 1985; pp. 572–581.

40. Murphy, L.; Simpson, A.; Dandy, G. *Pipe Network Optimization Using an Improved Genetic Algorithm*; Research Report; Rep.109; University of Adelaide, Department of Civil and Environmental Engineering: Adelaide, Australia, 1993.

41. Kim, J.; Kim, T.; Yoon, Y. A study on the pipe network system design using non-linear programming. *J. Korean Water Resour. Assoc.* **1994**, *27*, 59–67.

42. Walski, T.; Brill, E.; Gessler, J.; Goulter, I.; Jeppson, R.; Lansey, K.; Lee, H.; Liebman, J.; Mays, L.; Morgan, D.; et al. Battle of the network models: Epilogue. *J. Water Resour. Plan. Manag. ASCE* **1987**, *113*, 191–203.

43. Sherali, H.D.; Subramanian, S.; Loganathan, G. Effective Relaxations and Partitioning Schemes for Solving Water Distribution Network Design Problems to Global Optimality. *J. Glob. Optim.* **2001**, *19*, 1–26.

44. Fujiwara, O.; Khang, D.B. A two-phase decomposition method for optimal design of looped water distribution networks. *Water Resour. Res.* **1990**, *26*, 539–549.

45. Lee, S.C.; Lee, S.I. Genetic algorithms for optimal augmentation of water distribution networks. *J. Korean Water Resour. Assoc.* **2001**, *34*, 567–575.

46. Bragalli, C.; D'Ambrosio, C.; Lee, J.; Lodi, A.; Toth, P. On the optimal design of water distribution networks: A practical MINLP approach. *Optim. Eng.* **2012**, *13*, 219–246.

47. van Zyl, J.; Savic, D.; Walters, G. Operational optimization of water distribution systems using a hybrid genetic algorithm. *J. Water Resour. Plan. Manag.* **2004**, *130*, 160–170.

48. Ostfeld, A.; Uber, J.G.; Salomons, E.; Berry, J.W.; Hart, W.E.; Phillips, C.A.; Watson, J.P.; Dorini, G.; Jonkergouw, P.; Kapelan, Z.; et al. The Battle of the Water Sensor Networks (BWSN): A Design Challenge for Engineers and Algorithms. *J. Water Resour. Plan. Manag.* **2008**, *134*, 556–568.

49. Di Nardo, A.; Di Natale, M.; Giudicianni, C.; Laspidou, C.; Morlando, F.; Santonastaso, G.; Kofinas, D. Spectral analysis and topological and energy metrics for water network partitioning of Skiathos island. *Eur. Water* **2017**, *58*, 423–428.

50. Di Nardo, A.; Di Natale, M.; Gisonni, C. La distrettualizzazione delle reti idriche per il controllo delle pressioni (il sito pilota di Monteruscello). In Proceedings of the Atti del XXXI Convegno di Idraulica e Costruzioni Idrauliche, Perugia, Italy, 9–12 Settembre 2008.

51. Herrera, M.; Canu, S.; Karatzoglou, R.; Perez-Garcia, R.; Izquierdo, J. An approach to water supply clusters by semi-supervised learning. In Proceedings of the International Congress on Environmental Modelling and Software Modelling (iEMSs) 2010 for Environment Sake, Fifth Biennial Meeting, Ottawa, ON, Canada, 5–8 July 2010.

52. Marchi, A.; Salomons, E.; Ostfeld, A.; Kapelan, Z.; Simpson, A.R.; Zecchin, A.C.; Maier, H.R.; Wu, Z.Y.; Elsayed, S.M.; Song, Y.; et al. Battle of the Water Networks II. *J. Water Resour. Plan. Manag.* **2014**, *140*, 04014009.
53. Reca, J.; Martinez, J. Genetic algorithms for the design of looped irrigation water distribution networks. *Water Resour. Res.* **2006**, *42*, W05416.
54. Marchis, M.D.; Fontanazza, C.M.; Freni, G.; Loggia, G.L.; Napoli, E.; Notaro, V. A model of the filling process of an intermittent distribution network. *Urban Water J.* **2010**, *7*, 321–333.
55. Lippai, I. Colorado Springs Utilities Case Study: Water System Calibration/Optimization. In *Pipelines 2005: Optimizing Pipeline Design, Operations, and Maintenance in Today's Economy*; American Society of Civil Engineers: Reston, VA, USA, 2005.
56. Farmani, R.; Savic, D.A.; Walters, G.A. Evolutionary multi-objective optimization in water distribution network design. *Eng. Optim.* **2005**, *37*, 167–183.
57. Salomons, E.; Skulovich, O.; Ostfeld, A. Battle of Water Networks DMAs: Multistage Design Approach. *J. Water Resour. Plan. Manag.* **2017**, *143*, 04017059.
58. Erdos, P.; Renyi, A. On random graphs. *Publ. Math. Debr.* **1959**, *6*, 290–297.
59. Wormald, N.C. Generating random regular graphs. *J. Algorithms* **1984**, *5*, 247–280.
60. McKay, B.D.; Wormald, N.C. Uniform generation of random regular graphs of moderate degree. *J. Algorithms* **1990**, *11*, 52–67.
61. Giustolisi, O.; Simone, A.; Ridolfi, L. Network structure classification and features of water distribution systems. *Water Resour. Res.* **2017**, *53*, 3407–3423.
62. Pothen, A.; Simon, H.D.; Liou, K.P. Partitioning Sparse Matrices with Eigenvectors of Graphs. *SIAM J. Matrix Anal. Appl.* **1990**, *11*, 430–452.

![water logo] *water*

MDPI

Article

Comparing Topological Partitioning Methods for District Metered Areas in the Water Distribution Network

Haixing Liu [1], Mengke Zhao [1], Chi Zhang [1,*] and Guangtao Fu [2]

[1] School of Hydraulic Engineering, Dalian University of Technology, Dalian 116023, Liaoning, China; lhx_526@163.com (H.L.); dutzmk@163.com (M.Z.)

[2] Center for Water Systems, College of Engineering, Mathematics, and Physical Sciences, University of Exeter, North Park Rd., Exeter EX4 4QF, UK; g.fu@exeter.ac.uk

* Correspondence: czhang@dlut.edu.cn

Received: 13 February 2018; Accepted: 19 March 2018; Published: 23 March 2018

Abstract: This paper presents a comparative analysis of three partitioning methods, including Fast Greedy, Random Walk, and Metis, which are commonly used to establish the district metered areas (DMAs) in water distribution systems. The performance of the partitioning methods is compared using a spectrum of evaluation indicators, including modularity, conductance, density, expansion, cuts, and communication volume, which measure different topological characteristics of the complex network. A complex water distribution network EXNET is used for comparison considering two cases, i.e., unweighted and weighted edges, where the weights are represented by the demands. The results obtained from the case study network show that the Fast Greedy has a good overall performance. Random Walk can obtain the relative small cut edges, but severely sacrifice the balance of the partitions, in particular when the number of partitions is small. The Metis method has good performance on balancing the size of the clusters. The Fast Greedy method is more effective in the weighted graph partitioning. This study provides an insight for the application of the topology-based partitioning methods to establish district metered areas in a water distribution network.

Keywords: water distribution network; graph partitioning; modularity; district metered areas

---

1. Introduction

Districted metered areas (DMAs) play an essential role in water distribution system (WDS) management, such as pressure management, leakage reduction, and water quality incident control [1–4]. Many partitioning methods are available to divide the water distribution network into isolated DMAs, whose inlets and outlets can be monitored with flow and pressure meters. However, there is a lack of understanding in the performance of various partitioning methods, which will be investigated in this paper.

The partitioning methods have stemmed from, and advanced, the field of complex networks [5,6]. The water distribution network—a typical complex network—is usually constructed under the ground and along the road and provides drinking water to communities and cities. The network layouts are usually shaped by the community characteristics, such as geography and building distribution. The more developed the community is, the denser the connections (pipes) are in a community. The early research regarding network partitioning focuses on detecting the community structures [7,8]. This type of method is based on the modularity index, which is maximized using an optimization algorithm (e.g., greedy algorithm [9]) and able to quantify the network partitioning performance by the number of links and the degree of nodes [10].

Recent research has focused on the development of topology-based partitioning methods (e.g., graph theory, Metis, and Random Walk) to consider more network attributes. Most recently, Perelman et al. [3] developed a network partitioning approach from a practical perspective, in which the distribution network is partitioned based on the backbone transmission pipes identified in the first step. The Metis partitioning method [11] was used to divide the network topology into DMAs. Moreover, the water sources should be regarded as the main control elements for network partitioning [12] so that the source supply areas have the least influence on the hydraulic performance of the isolated zones and the finer communities are further divided by a graph partitioning method. Wright et al. [13] introduced the concept of dynamically-controlled DMAs for burst identification and leakage estimation, and it can be used to improve system resilience to permanent valve closures. The Walktrap algorithm, which is also referred to as Random Walk [14,15], was used to trap the walker in the dense cluster, and partition the water distribution network accordingly [16]. The aforementioned studies are the straightforward applications to network partitioning, however, the mechanism, characteristic, and applicability of the graphical partitioning methods (i.e., Metis, Random Walk) are not properly addressed. Moreover, there is no comprehensive comparison of the various topology partitioning algorithms and, therefore, it is difficult to determine which algorithm is more suitable and, thus, should be used for WDS partitioning in practice.

Network partitioning can be assessed using hydraulic indicators and topological indicators. Additionally, economic criteria can also be used to consider the costs and benefits involved by the reconstruction and installation of partitioning but the difficulty lies in data availability. The hydraulic indicators used in the literature include resilience index, water age, entropy, and pressure-related performance index [12], however, this involves the use of a hydraulic model for assessment [17]. Topological indicators are normally easy for computation and use, thus, a spectrum of topological indicators are used in this study, including the most commonly used modularity index.

This paper aims to compare three widely used partitioning methods, including Fast Greedy [9], Random Walk [15], and Metis [18], using a spectrum of topology-based indicators. Note that heuristic partitioning methods are also popular in water network partitioning [19], however, they are not investigated in this study since they are more customized for specific water networks with problem domain knowledge required. In comparison, graphical partitioning methods (i.e., Fast Greedy, Random Walk, and Metis), developed from fundamental graphical theories, are easier to use and can be applied to any WDS networks, thus, they are investigated in this study for network partitioning. A water distribution network, EXNET [20], is investigated with two cases: weighted and unweighted graphs. The advantage and disadvantage of the partitioning algorithms are discussed to support their application to the water distribution network.

2. Partitioning Methods

A complex network can be represented by an undirected graph $G(V, E)$, where V is a set of vertices which indicates the nodes in the network, and E notes a set of edges, representing pipes and other link elements in a hydraulic model, such as valves and pumps. *Cluster*, which represents a group of edges and vertices, is commonly used in the graph analysis, while *community* in the water distribution network is the cluster of pipes and nodes. In this paper, the terms—cluster and community—are used interchangeably.

2.1. Fast Greedy

The Fast Greedy partitioning method, developed by Clauset et al. [9], is a modularity-based topology analysis method. The modularity is a measure of the strength of the network partitioning to clusters (i.e., communities). It is based on a modularity function, which is optimized to find the best cluster of a water distribution network. The modularity function is formulated as:

$$M = \frac{1}{2m} \sum_{vw} \left[A_{vw} - \frac{k_v k_w}{2m} \right] \delta(c_v, c_w) \tag{1}$$

where M is the modularity value; m is the number of edges in the graph; A_{vw} is an element of adjacency matrix of the network, $A_{vw} = 1$ when vertices v and w are connected, otherwise $A_{vw} = 0$; k_v is the degree of a vertex v and is defined as the number of edges connected to it, $k_v = \sum_w A_{vw}$. c_v and c_w are the identifiers of a network cluster. δ is the function of the summation of the same clusters (if $c_v = c_w$, then δ = 1, and otherwise $\delta = 0$). Maximizing M implies maximization of the intra-links in a community and minimization of the inter-links between communities. The higher value of the modularity, the better the solution of the partitioning in the network.

The Fast Greedy partitioning method uses a greedy optimization approach to maximize the modularity function, as shown in Figure 1. Each vertex represents a single community at the initial step. Then it attempts to combine any two linking communities together, with an aim to maximizing the modularity value. The combination process is repeated until the network merges to one community. The Fast Greedy partitioning method improves its computational efficiency through storing only those pairs of communities that are linked by one or more edges, instead of manipulating the entire sparse matrix. The new matrix with the efficient data structure tracks the linking communities that can potentially increase in the modularity value. Therefore, the computational time and memory can be substantially reduced. A more detailed description with respect to the Fast Greedy method can be found in the studies by Clauset et al. [9], Vincent et al. [5], and Fortunato [21].

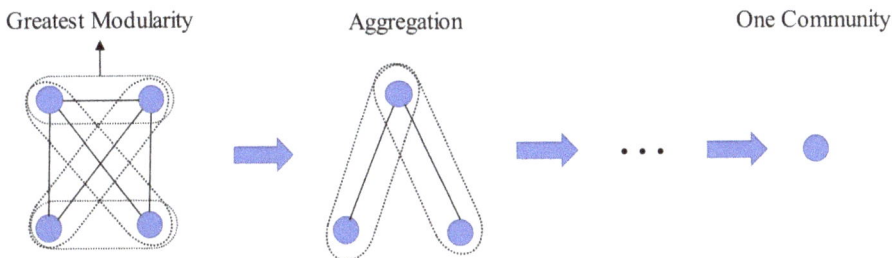

Figure 1. Modularity-based aggregation in Fast Greedy.

2.2. Random Walk

The Random Walk partitioning method relies on the fact that the walker can be retained in a part of densely-connected networks (i.e., a cluster) as much as possible before leaving for another cluster. An infinite random walk is quantified by the map equation, which is the average number of bits per step in the clusters of the network [14]:

$$L(M) = q_{\frown} H(\mathcal{Q}) + \sum_{i=1}^{m} p_{\circlearrowleft}^i H\left(\mathcal{P}^i\right) \tag{2}$$

where $q_{\frown}$ represents the probability of the walker switching the clusters per step, $q_{\frown} = \sum_{i=1}^{m} q_{i\frown}$. $q_{i\frown}$ is the probability of the walk running out of the cluster i per step. $p_{\circlearrowleft}^i$ is calculated to weight the movements within cluster i, $p_{\circlearrowleft}^i = q_{i\frown} + \sum_{\alpha \in i} p_{\alpha}$, where α is the node in the cluster i and p_{α} is the visit frequency at node α. $H(\mathcal{Q})$ and $H(\mathcal{P}^i)$ are the probability-weighted average length of the walk in the network and in the cluster i, respectively. $H(\mathcal{Q})$ and $H(\mathcal{P}^i)$ are quantified by Shannon' entropy [22]. $H(\mathcal{Q})$ is the entropy of movements between the clusters:

$$H(\mathcal{Q}) = \sum_{i=1}^{m} \frac{q_{i\curvearrowright}}{\sum_{j=1}^{m} q_{i\curvearrowright}} \log\left(\frac{q_{i\curvearrowright}}{\sum_{j=1}^{m} q_{i\curvearrowright}}\right) \tag{3}$$

The entropy of movements within cluster i is:

$$H(\mathcal{P}^i) = \frac{q_{i\curvearrowright}}{q_{i\curvearrowright} + \sum_{\beta \in i} p_\beta} \log\left(\frac{q_{i\curvearrowright}}{q_{i\curvearrowright} + \sum_{\beta \in i} p_\beta}\right) + \sum_{\alpha \in i} \frac{p_\alpha}{q_{i\curvearrowright} + \sum_{\beta \in i} p_\beta} \log\left(\frac{p_\alpha}{q_{i\curvearrowright} + \sum_{\beta \in i} p_\beta}\right) \tag{4}$$

The random walk presents that the walker (i.e., a random moving point) is able to linger in a relatively isolated area (i.e., with dense links) longer than the sparse network. That is because, in dense communities, there are fewer routes to move outside of the communities, which matches the concept of the network partitioning. The map equation (Equation (2)) has two terms to represent the average numbers of bits to quantify the movements in the clusters and between clusters, respectively. The variables of node visit probability p_α and the leaving probability $q_{i\curvearrowright}$ in the map equation are updated in each step during an optimization process. The smaller value of the map equation is desirable and, thus, the minimization problem associated with map equation is formulated. The problem is solved by a deterministic greedy search algorithm and then the solutions are refined (i.e., the description length of the walk in the clusters) by the simulated annealing algorithm. More details regarding the random walk method can be found in the follow studies [15,23,24].

2.3. Metis

The Metis partitioning method [18] uses a k-way (i.e., k clusters) partitioning method to minimize the number of cuts (i.e., links between two clusters), while maintaining the balance of the clusters. The partitioning process consists of coarsening, partitioning, and uncoarsening phases, as explained below.

(1) *Coarsening Phase.* The coarsening phase is to combine the incident vertices and form an updated graph that includes a smaller number of vertices. The modified Heavy Edge Matching (mHEM) method [18] is implemented to decrease the average degree of coarser graphs by finding the edges with the maximum weights. Hence, the total weight of the edges in a coarser graph is reduced by the maximal matching.

(2) *Partitioning Phase.* This phase is realized by a high-quality bisection process in which the graph is divided into two parts of approximately equal weights. The Kernighan-Lin algorithm [25] is used in the partitioning phase. In the Kernighan-Lin algorithm, the vertices on the boundary of either partitions attempting to be swapped to obtain a smaller number of edge cuts. If the swapping is accepted, the new clusters are formed and this process continues until no more improvement on the uniformity of clusters. Since the partitioning is implemented in a lumped graph after the coarsening phase, the time consumed by the partitioning is much less than in the original graph.

(3) *Uncoarsening Phase.* Based on the partitioning results obtained from the second phase, the coarser clusters are gradually projected back to the original graph. Due to the relative good partitioning results derived, the refinement phase is only implemented on the small number of vertices that link the different clusters. The modified Global Kernighan-Lin Refinement is proposed by Karypis and Kumar [18], which enhances the capability of the partitioning refinement in order to escape the local optima.

The Metis method uses several heuristic approaches (i.e., swapping vertices and bisection method) to deal with the graph partitioning in a three-phase process. The heuristic that swaps the vertices at the rim of the clusters can save the computational time and is able to derive the good clusters.

The swapping action is beneficial for balancing the adjacent the clusters in a straightforward way. These heuristics can improve the computational efficiency, avoiding manipulating a large scale matrix. The heuristic (swapping boundary nodes) is not a global search approach, hence, the resulting partitioning is not necessarily the global optimal solution. The detailed description of Metis can be found in the following studies [18,26].

3. Evaluation Indicators

The clusters derived from the topological partitioning methods above need to be evaluated to determine which method can derive a better WDS partitioning solution. Perelman et al. [27] recommended to visually show the graphics of partitioning solutions, since the partitioning results and the linkage between communities can be seen intuitively. This helps the decision maker to understand the partitioning schemes and thus make an informed decision. Additionally, quantitative evaluation is usually conducted to examine the effect of the partitioning [12,28,29]. In this paper, several indicators in the context of topological analysis are proposed to quantify the performance of partitioning methods.

Modularity is commonly used for DMA partitioning [2,30,31], which gives a measure of the strength of network partitioning into modules (i.e., communities). The modularity indicator indicates the density of pipes in a community and the linkage between the communities. Therefore, the greater the modularity value, the more community-like the network. The modularity value is calculated using Equation (1).

Conductance represents the fraction of the total pipes that link to other pipes outside the community. The formula of the conductance indicator is given as:

$$C = \frac{c_S}{2m_S + c_S} \tag{5}$$

where c_S is the number of edges on the boundary of a cluster, and m_S is the number of edges inside the cluster S. These two variables are calculated as:

$$m_S = |\{(u,v); u \in S, v \in S\}| \tag{6}$$

$$c_S = |\{(u,v); u \in S, v \notin S\}| \tag{7}$$

The conductance indicator for a given DMA is calculated for the incident pipes that link to other DMAs. The smaller the conductance value, the fewer the pipes link outside, thus, network partitions are more community-like.

Density is defined as the intra-cluster density, $D = \frac{m_S}{n_S(n_S-1)/2}$ where n_S is the number of nodes in S, $n_S = |S|$. The density indicator delivers the information of the pipes' sparseness in a cluster. It is the ratio of the number of edges in the cluster to maximal number of possible edges. The greater the density value, the higher the density of pipes inside the cluster, thus, the better the partitioning method.

Expansion measures the number of pipes per node that link outside the cluster. The expansion indicator is given as $E = \frac{c_S}{n_S}$, where n_S is the number of nodes in S. The smaller the expansion value, the less links expand to the outside clusters, thus, the network partitioning is preferred.

Cuts indicates the average number of pipes by which a cluster links to the other. This indicator denotes the number of valves or meters need to be installed for partitioning. The fewer number of cuts, the partitioning retrofit is more cost-saving.

Communication volume (CV) is a measure of the communication complexity amongst the clusters. V_b is the nodes at the cluster boundary, i.e., linking to other cluster(s). For each node $v \in V_b$, CV is the number of outside clusters that v is adjacent to. The CV value is smaller, the communication cost is lower, thus, the partitioning method is better.

4. Case Study

The topological partition methods are applied to a large-scale water network: EXNET, as shown in Figure 2. The network serves a population of 400,000. The network system consists of 2416 pipes and 1891 nodes. The elevated reservoirs provide the energy to the entire system.

The weights could be allocated to edges or vertices for graph partitioning. In this paper, the demand distributed along a pipe is regarded as the edge weight, since the demand is highly related to the community scale. The distribution demand of a pipe is calculated: $w_d = \sum_{i=1}^{k} \frac{1}{n_i} D_i$, where D_i is the demand at node i, n_i is the number of pipes that link to node i, and k represents the number of nodes of a pipe and it normally equals 1 or 2. Therefore, in this study, the weighted and unweighted graphs are comparatively investigated in partitioning.

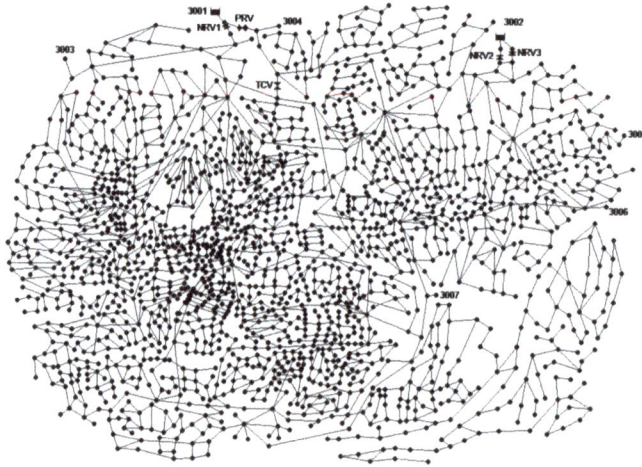

Figure 2. EXNET water distribution network.

5. Results and Discussion

5.1. Unweighted Partitioning

The EXNET network is partitioned into 5, 10, 15, 20, 30, and 50 communities using the Fast Greedy, Random Walk, and Metis algorithms, respectively. The indicators are calculated for each partitioning in the case study network. The averaged indicator values of all communities in the network are shown in Figure 3. The modularity indicator varies significantly when the number of communities is less than 20 in this case, but has little variation in the cases of more than 20 communities. Comparing the three methods, the solutions derived from the Fast Greedy method are better than those from the other two methods, though they are very close to those from the Random Walk method. In terms of the density results, Metis outperforms the Random Walk and Fast Greedy methods. With respect to the conductance and expansion indicators, Fast Greedy and Random Walk are better than Metis. However, Random Walk produces the smaller number of cuts and communication volume than the other methods. In all, the three methods have their respective advantages in terms of the different indicators (i.e., modularity, conductance, density, expansion, cut, and communication volume). It is impossible to say one method is absolutely better than others when all indicators are considered.

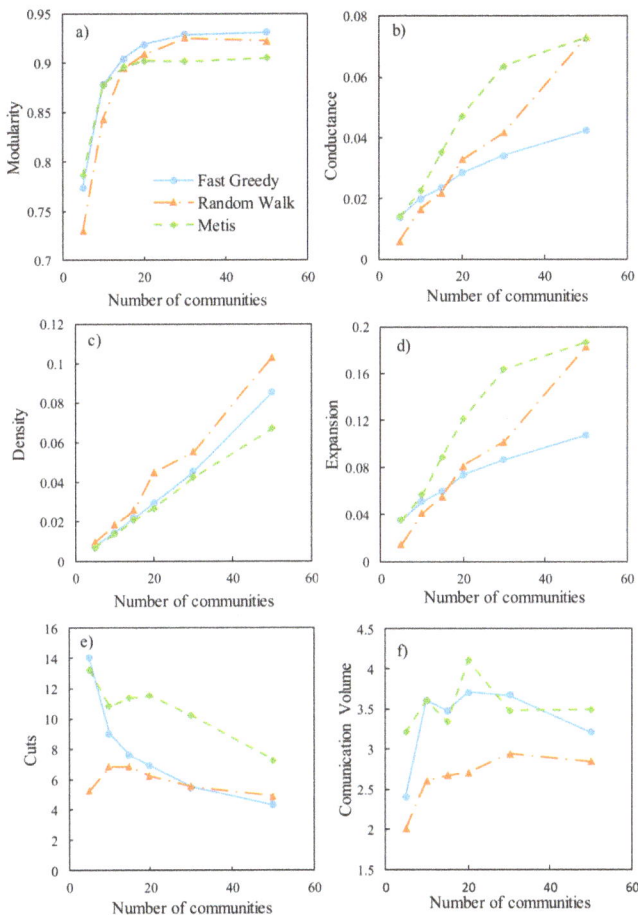

Figure 3. The indicator values of the three partitioning methods. (**a**) Modularity; (**b**) Conductance; (**c**) Density; (**d**) Expansion; (**e**) Cuts; and (**f**) Communication Volume.

The numbers of nodes in all communities of the network are shown in Figure 4. The maximum (top bar), the minimum (bottom bar), the 25 and 75 percentiles (box top and bottom edges), the median (line in the box), and the mean (cross) are demonstrated in Figure 4. Usually, the similar numbers of nodes throughout all communities are desired as a result of DMA partitioning. As can be seen in Figure 4, Metis shows a smaller variation of the node number and, thus, its partitioning result outperforms the Fast Greedy and Random Walk from the uniformity perspective. Random Walk shows a larger variation in node number, particularly when the cluster number is small. However, when the cluster number becomes large, e.g., 30 and 50 clusters, the variations derived from Fast Greedy and Random Walk are very similar, but are still larger than that from Metis. Metis includes the outstanding feature on the node distribution of partitioning, as it has two in three phases (i.e., partitioning and refinement) to adjust the balance of the size of communities. If the uniformity of nodes in the communities is pursued in the network partitioning, the Metis algorithm is strongly recommended.

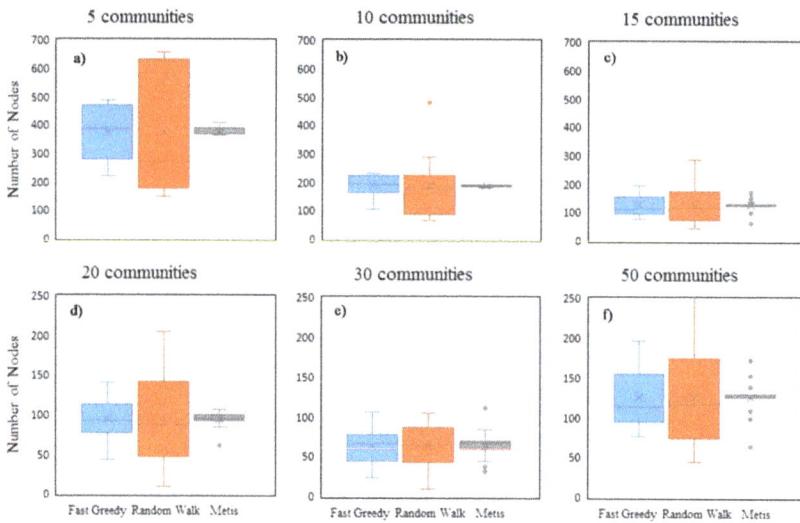

Figure 4. The number of nodes in the communities from three partitioning methods. Divided into (**a**) 5 communities; (**b**) 10 communities; (**c**) 15 communities; (**d**) 20 communities; (**e**) 30 communities, and (**f**) 50 communities.

5.2. Weighted Partitioning

The demands are used as weights to differentiate the crucial edges. In doing so, the crucial edges (i.e., with large weights) are not prone to be chosen as the cut edges (the links between communities), since flow meters and isolation valves are normally installed at the boundary of communities (i.e., cut edges). The results of the indicators are shown in Figure 5. As can be seen in Figure 5, the indicator values are substantially different from those of the unweighted partitioning. In particular, the Random Walk method has the distinct performance in comparison with Metis and Fast Greedy. The Random Walk shows the lower modularity than Fast Greedy and Metis, which indicates that the overall performance of Random Walk is worse, with the only exception being the case of 50 communities. Fast Greedy obtained the higher modularity values in both weighted and unweighted network partitioning, since the Fast Greedy optimizes the modularity function directly. Similar to the unweighted partitioning, the cuts and communication volume of Random Walk are better than the other two methods.

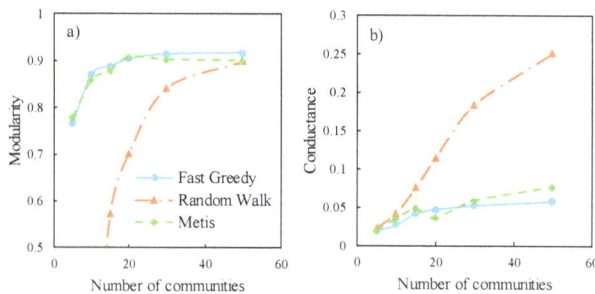

Figure 5. *Cont.*

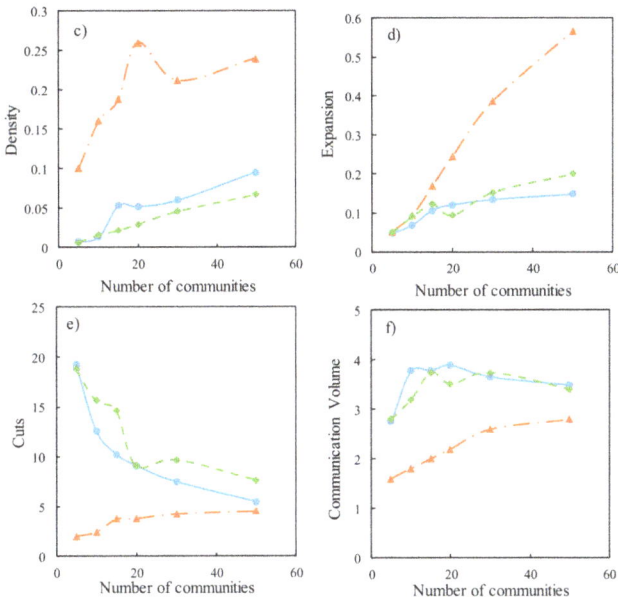

Figure 5. The indicators of the three partitioning methods in the weighted network. (**a**) Modularity; (**b**) Conductance; (**c**) Density; (**d**) Expansion; (**e**) Cuts; and (**f**) Communication Volume.

Figure 6 shows the demand statistics in all communities with and without weights in partitioning. The Fast Greedy method shows the best performance, since the demands become more condensed (i.e., demands for all clusters concentrate on the average) when taking the weights into account. The Metis shows the inconsistent performance, i.e., good performance at five communities' partitioning, but poor performance at 20 and 30 communities' partitioning. The Random Walk method seems incapable of dealing with the weighted partitioning.

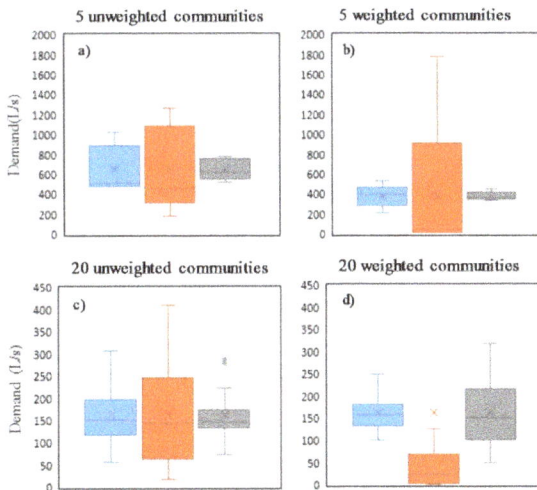

Figure 6. *Cont.*

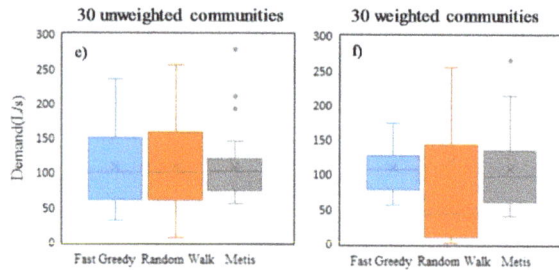

Figure 6. The demands of all communities from the three partitioning methods with weighted and unweighted graphs. (**a**) 5 communities, (**c**) 20 communities, (**e**) 30 communities in unweighted graph; and (**b**) 5 communities, (**d**) 20 communities, (**f**) 30 communities in weighted graph.

5.3. Discussion

Random Walk is mainly concerned with the cut number, but significantly sacrifices the balance of network partitions. In some engineering fields, including water distribution systems, balanced network partitioning is vital. Hence, Metis and Fast Greedy methods are preferable for water distribution network partitioning. Random Walk is applied in the directed and weighted graph partitioning in Rosvall and Bergstrom [15]. That might be the reason why it does not perform well in the water distribution network that is an undirected graph.

The EXNET is a water distribution network that is modified from a real network in the UK. When applied to real-world networks, the partitioning methods need to consider the following factors: (1) the real-world WDS has more complex structure and components (pump, tank, reservoir) than the model network; (2) some small pipes of network branches in a complex network could be removed because the focus should be on the main pipes of the real-world network regarding network partitioning; and (3) hydraulic isolation (valve operation) should be considered in partitioning.

Further, network partitioning, in practice, should concern more influence factors that might significantly impact the partitioning results, for example, pipe length, pressure area, water quality deterioration, minimum night flow, degree of demand variation, and fire flow capacity, to name a few [12,19].

6. Conclusions

This paper evaluated three partitioning methods, including Fast Greedy, Random Walk, and Metis, using a spectrum of topological evaluation indicators. The three partitioning methods are commonly used for water network partitioning and are developed from different theories. A large-scale water distribution network, i.e., EXNET, is used for comparison analysis. The unweighted and weighted graphs (represented by demands) are comparatively analyzed in partitioning. According to the results obtained, key conclusions are given below:

(1) The Fast Greedy method is better than the Random Walk and Metis methods in terms of the modularity indicator for a single test case. Fast Greedy directly optimizes the modularity function which measures the network partitioning in a comprehensive way. The good overall performance of Fast Greedy tends to weaken inter-community links and strengthen intra-community links. Random Walk results in a smaller number of edge cuts, which indicates the connections between the communities are not strong; conversely, Fast Greedy and Metis lead to the dense links in the communities.

(2) Fast Greedy is able to reflect the effect of weighted edges in the partitioning, while Random Walk does not perform well in the weighted graph partitioning. Metis shows the varied performance with respect to the weighted graph, and its performance depends on the number of communities.

(3) Metis is able to obtain well-balanced communities in terms of a similar number of nodes across communities. This feature is preferred in the water distribution network partitioning, since the uniformity of DMAs is convenient for the management and maintenance tasks in the utilities.

This study provides an insight into the partitioning methods from a topological perspective. The three methods used in the study are all developed in the context of complex networks, thus, they should be tailored when applied to DMAs partitioning. The mechanism, characteristic, and applicability of the three partitioning algorithms demonstrated in this paper would provide useful insight to extend the applications of network partitioning.

Only one water distribution network (EXNET) is applied in this study, and more case studies should be investigated based on the methodology in future work. The future work should investigate how to establish the suitable valves and meters to form the relatively isolated zones and consider the hydraulic and economic performance when assessing network partitioning.

Acknowledgments: The work is financially supported by the National Key Research and Development Program (2017YFC0406005) and the National Natural Science Foundation of China (91547116 and 51708086). This work is also partially funded by the UK Engineering and Physical Sciences Research Council (EPSRC) project BRIM (EP/N010329/1), and the China Postdoctoral Science Foundation (2016M601309).

Author Contributions: All authors contributed to the original idea for this paper; Haixing Liu performed the methods to the case and analyzed the data; Mengke Zhao edited the formulas and made the figures for the manuscript; Haixing Liu and Mengke Zhao wrote the draft of manuscript; Chi Zhang and Guangtao Fu revised the draft into the final manuscript.

Conflicts of Interest: The authors declare no conflict of interest.

References

1. Alvisi, S.; Franchini, M. A heuristic procedure for the automatic creation of district metered areas in water distribution systems. *Urban Water J.* **2014**, *11*, 137–159. [CrossRef]
2. Giustolisi, O.; Ridolfi, L. A novel infrastructure modularity index for the segmentation of water distribution networks. *Water Resour. Res.* **2014**, *50*, 7648–7661. [CrossRef]
3. Perelman, L.S.; Allen, M.; Preis, A.; Iqbal, M.; Whittle, A.J. Flexible reconfiguration of existing urban water infrastructure systems. *Environ. Sci. Technol.* **2015**, *49*, 13378–13384. [CrossRef] [PubMed]
4. Di Cristo, C.; Leopardi, A.; de Marinis, G. Assessing measurement uncertainty on trihalomethanes prediction through kinetic models in water supply systems. *J. Water Suppl.* **2015**, *64*, 516–528. [CrossRef]
5. Vincent, D.B.; Jean-Loup, G.; Renaud, L.; Etienne, L. Fast unfolding of communities in large networks. *J. Stat. Mech.* **2008**, *2008*, P10008.
6. Lancichinetti, A.; Fortunato, S. Community detection algorithms: A comparative analysis. *Phys. Rev. E* **2009**, *80*, 056117. [CrossRef] [PubMed]
7. Di Nardo, A.; Di Natale, M.; Santonastaso, G.; Tzatchkov, V.; Alcocer-Yamanaka, V. Water network sectorization based on graph theory and energy performance indices. *J. Water Resour. Plan. Manag.* **2013**, *140*, 620–629. [CrossRef]
8. Diao, K.; Fu, G.; Farmani, R.; Guidolin, M.; Butler, D. Twin-hierarchy decomposition for optimal design of water distribution systems. *J. Water Resour. Plan. Manag.* **2015**, *142*, C4015008. [CrossRef]
9. Clauset, A.; Newman, M.E.J.; Moore, C. Finding community structure in very large networks. *Phys. Rev. E* **2004**, *70*, 066111. [CrossRef] [PubMed]
10. Giustolisi, O.; Ridolfi, L.; Berardi, L. General metrics for segmenting infrastructure networks. *J. Hydroinform.* **2015**, *17*, 505–517. [CrossRef]
11. Karypis, G.; Kumar, V. *Metis: A Software Package for Partitioning Unstructured Graphs, Partitioning Meshes, and Computing Fill-Reducing Orderings of Sparse Matrices*; Department of Computer Science and Engineering, University of Minnesota, Army HPC Research Center: Minneapolis, MN, USA, 2013.
12. Scarpa, F.; Lobba, A.; Becciu, G. Elementary DMA design of looped water distribution networks with multiple sources. *J. Water Resour. Plan. Manag.* **2016**, *142*, 04016011. [CrossRef]

13. Wright, R.; Abraham, E.; Parpas, P.; Stoianov, I. Control of water distribution networks with dynamic DMA topology using strictly feasible sequential convex programming. *Water Resour. Res.* **2015**, *51*, 9925–9941. [CrossRef]

14. Pons, P.; Latapy, M. Computing communities in large networks using random walks. In Proceedings of the 2005 20th International Symposium Computer and Information Sciences, Istanbul, Turkey, 26–28 October 2005; Yolum, P., Güngör, T., Gürgen, F., Özturan, C., Eds.; Springer: Berlin/Heidelberg, Germany, 2005; pp. 284–293.

15. Rosvall, M.; Bergstrom, C.T. Maps of random walks on complex networks reveal community structure. *Proc. Natl. Acad. Sci. USA* **2008**, *105*, 1118–1123. [CrossRef] [PubMed]

16. Campbell, E.; Izquierdo, J.; Montalvo, I.; Ilaya-Ayza, A.; Pérez-García, R.; Tavera, M. A flexible methodology to sectorize water supply networks based on social network theory concepts and multi-objective optimization. *J. Hydroinform.* **2016**, *18*, 62–76. [CrossRef]

17. Liu, H.; Walski, T.; Fu, G.; Zhang, C. Failure impact analysis of isolation valves in a water distribution network. *J. Water Resour. Plan. Manag.* **2017**, *143*, 04017019. [CrossRef]

18. Karypis, G.; Kumar, V. Multilevelk-way partitioning scheme for irregular graphs. *J. Parallel Distrib. Comput.* **1998**, *48*, 96–129. [CrossRef]

19. Di Nardo, A.; Giudicianni, C.; Greco, R.; Herrera, M.; Santonastaso, G. Applications of graph spectral techniques to water distribution network management. *Water* **2018**, *10*, 45. [CrossRef]

20. Farmani, R.; Savic, D.A.; Walters, G.A. Evolutionary multi-objective optimization in water distribution network design. *Eng. Optim.* **2005**, *37*, 167–183. [CrossRef]

21. Fortunato, S. Community detection in graphs. *Phys. Rep.* **2010**, *486*, 75–174. [CrossRef]

22. Shannon, C.E. A mathematical theory of communication. *Bell Syst. Tech. J.* **1948**, *24*, 379–423. [CrossRef]

23. Rosvall, M.; Axelsson, D.; Bergstrom, C.T. The map equation. *Eur. Phys. J. Spec. Top.* **2009**, *178*, 13–23. [CrossRef]

24. Rosvall, M.; Esquivel, A.V.; Lancichinetti, A.; West, J.D.; Lambiotte, R. Memory in network flows and its effects on spreading dynamics and community detection. *Nat. Commun.* **2014**, *5*, 4630. [CrossRef] [PubMed]

25. Kernighan, B.W.; Lin, S. An efficient heuristic procedure for partitioning graphs. *Bell Syst. Tech. J.* **1970**, *49*, 291–307. [CrossRef]

26. Karypis, G.; Kumar, V. Multilevel algorithms for multi-constraint graph partitioning. In Proceedings of the 1998 ACM/IEEE Conference on Supercomputing, San Jose, CA, USA, 7–13 November 1998.

27. Sela Perelman, L.; Allen, M.; Preis, A.; Iqbal, M.; Whittle, A.J. Automated sub-zoning of water distribution systems. *Environ. Model. Softw.* **2015**, *65*, 1–14. [CrossRef]

28. Alvisi, S. A new procedure for optimal design of district metered areas based on the multilevel balancing and refinement algorithm. *Water Resour. Manag.* **2015**, *29*, 4397–4409. [CrossRef]

29. Herrera, M.; Abraham, E.; Stoianov, I. A graph-theoretic framework for assessing the resilience of sectorised water distribution networks. *Water Resour. Manag.* **2016**, *30*, 1685–1699. [CrossRef]

30. Diao, K.; Zhou, Y.; Rauch, W. Automated creation of district metered area boundaries in water distribution systems. *J. Water Resour. Plan. Manag.* **2012**, *139*, 184–190. [CrossRef]

31. Giustolisi, O.; Ridolfi, L. New modularity-based approach to segmentation of water distribution networks. *J. Hydraul. Eng.* **2014**, *140*, 04014049. [CrossRef]

water

MDPI

Article

Vulnerability Assessment to Trihalomethane Exposure in Water Distribution Systems

Claudia Quintiliani [1,2,*], **Cristiana Di Cristo** [1] **and Angelo Leopardi** [1]

[1] Department of Civil and Mechanical Engineering, University of Cassino and Southern Lazio, 03043 Cassino, Italy; dicristo@unicas.it (C.D.C.); a.leopardi@unicas.it (A.L.)

[2] Department of Civil Engineering and Architecture, University of Pavia, 27100 Pavia, Italy

* Correspondence: claudia.quintiliani@unipv.it or c.quintiliani@unicas.it; Tel.: +39-334-341-0903

Received: 28 May 2018; Accepted: 6 July 2018; Published: 10 July 2018

Abstract: Chlorination is an effective and cheap disinfectant for preventing waterborne diseases-causing microorganisms, but its compounds tend to react with the natural organic matter (NOM), forming potentially harmful and unwanted disinfection by-products (DBPs) such as trihalomethanes (THMs), haloacetic acids (HAAs), and others. The present paper proposes a methodology for estimating the vulnerability with respect to users' exposure to DPBs in water distribution systems (WDSs). The presented application considers total THMs (TTHMs) concentration, but the methodology can be used also for other types of DPBs. Five vulnerability indexes are adopted that furnish different kinds of information about the exposure. The methodology is applied to five case studies, and the results suggest that the introduced indexes identify different critical areas in respect to elevated concentrations of TTHMs. In this way, the use of the proposed methodology allows identifying the higher risk nodes with respect to the different kinds of exposure, whether it is a short period of exposure to high TTHMs values, or chronic exposure to low concentrations. The application of the methodology furnishes useful information for an optimal WDS management, for planning system modifications and district sectorization taking into account water quality.

Keywords: water distribution system; water quality; disinfection by-products; vulnerability

1. Introduction

The disinfection of drinking water increases users' protection by reducing their risk of exposure to pathogenic contamination and limiting the proliferation of microbial species [1,2]. Among the various technologies available nowadays, chlorination is still the major worldwide disinfection strategy, due mainly to its effectiveness [3,4], low cost, and simplicity (easy to produce, store, transport, and use).

Despite the proved effectiveness of chlorine in preventing waterborne disease-causing microorganisms, its compounds tend to react with the natural organic matter (NOM) naturally occurring in the resource through reactions in bulk water or close to the pipe wall, unintentionally forming potentially harmful and unwanted disinfection by-products (DBPs). It has been observed that some DBPs have possible adverse health effects. Indeed, they are carcinogenic potential substances [5–7], and some epidemiologic studies have raised the issue of potential adverse reproductive effects [8–10]. The exposure via non-ingestion routes, such as inhalation and dermal contact, may also pose risks to human health [11,12]. For this reason, a good water distribution management with respect to chlorination requires the solution of two conflicting objectives: to maintain sufficiently high chlorine residual and low enough DPBs concentrations in the whole system.

In the presented study, a methodology to assess the vulnerability respect to the DBPs exposure in a water distribution system (WDS) is proposed. The vulnerability assessment is useful for an optimal system operation for many purposes, such as for the optimization of chlorine booster functioning, to design public health strategies to reduce risks [13,14], for the optimal position of water quality monitoring stations [15,16], to obtain indications for epidemiological investigations [17], or for the district sectorization and skeletonization of a network based on water quality considerations [18,19].

Over the last 30 years, numerous DBPs have been studied and classified, including trihalomethanes (THMs), haloacetic acids (HAAs), haloacetonitriles (HANs), and haloketones (HKs). Even if it is unclear which specific DBPs are responsible for the adverse health effects [20,21], only two classes of DBPs—THMs and HAA—are regulated. The four THMs species (chloroform, bromodichloromethane, dibromochloromethane, and bromoform) are produced from sodium hypochlorite reacting with the NOM. Even if the total THMs (TTHMs) concentration does not give information on the HAA levels and it does not consider the relative role of the four THMs species in contributing to adverse health outcomes, it is often used as a surrogate for the measurement of DBPs [22].

Guidelines such as those from the World Health Organization [23] suggest maintaining in the system a free chlorine residual concentration in the range of 0.2–0.5 mg/L, while the TTHMs concentration is bounded by an upper regulation limit that is different in each country. For example, the United States (US) standard for the TTHMs concentration is 80 µg/L, the Canadian [24] and the Australian–New Zealand [25] guidelines set the maximum value to 100 µg/L and 250 µg/L, respectively, while in Italy, the regulation is very restrictive, with the limit value equal to 30 µg/L.

In the presented application, the methodology is performed considered the TTHMs concentration, furnishing indications to water companies for respecting the regulation limit. However, by adopting the adequate formation model, it can be used for studying other DPBs.

In the existing literature, water age (represented by residence time of the water in the system), contaminants, chlorine, and DBPs concentrations (such as bromide, TTHMs, and HAAs) are the main elements considered to evaluate indexes representative of a network vulnerability in terms of water quality.

Many studies estimated the exposure of the users to contaminants using different indexes. For example, Propato and Uber and Murrey et al. [26,27] quantified the contaminated water ingested by individuals during an event, evaluating the volume of polluted water that was delivered to the users. Nilson et al. [28] considered as a vulnerability index the total mass contaminated load in each node, and performed the sensitivity of the nodal mass distribution to various system characteristics and stochastic demands representation. Other studies have considered the impacts of contamination events without quantifying the ingestion by individuals, such as in Khanal et al. [29], which introduced an exposure index based on the percentage of the users exposed. Davis and Janke [30] accounted for the influence of ingestion timing in addition to the volume, and in a second work [31], the same authors also presented a probabilistic model for timing ingestion based on data collected by the American Time Use Survey in order to advise more realistic exposures.

Thompson et al. [32] estimated the vulnerability of a demand node in terms of exposure to low residual chlorine concentration, while Baoyu et al. [33] presented a vulnerability assessment model that also took into account the impact of water age, as well as the uncertainty related to bulk and wall reaction coefficients, demand multipliers, and pipe roughness.

Other works performed vulnerability analyses to identify the areas of the system and the population with a higher risk respect to DBPs exposure. More recently, Islam et al. [34] used a non-compliance potential index predicted using TTHMs, haloacetic acids (HAAs), and free residual chlorine as variables. Successively, Islam et al. [35] presented an index-based approach to locate chlorine booster stations in WDSs. The maximization of a Water Quality Index (WQI) based on the formation of TTHMs is used as objective of the optimization problem, and the required number of the stations is recommended based on a trade-off analysis of risk potentials, WQI, and the life cycle cost of a booster.

To perform the vulnerability analysis with respect to the consumers' exposure to TTHMs, it is fundamental to use models that are able to predict their formation within WDSs. Many mathematical approaches have been proposed that have been obtained by means of laboratory and field scaled experiments on raw and pre-treated water [11,36]. They usually account for the correlations existing between the formation of TTHMs and the elements that influence them, such as precursors, operational parameters (pH, contact time), and environmental conditions (temperature and season variability). Many predicting models are represented by empirical relationships between TTHMs concentrations and the main parameters affecting the TTHMs formation [37,38], while others are based on kinetic formulation that are, involved during chlorine bulk and wall reactions [39,40].

The present paper proposes a methodology for estimating the vulnerability respect to the TTHMs exposure in a WDS. Five vulnerability indexes furnishing different kinds of information about the exposure are adopted. The individually or aggregated use of these indexes allow identifying critical points in a WDS with respect to different kinds of exposure: high concentration exposure for a short period, or chronic exposure to low TTHMs concentrations, which are both important. The methodology is easy to apply and is useful for the water companies to evaluate the more critical zone in terms of water quality and also for analyzing future scenarios.

In this paper, the preliminary analysis performed on a literature network in Di Cristo et al. [41] is completed with further investigation, and extended to two more literature schemes and two more real water distribution systems.

The paper is structured in the following way: the following paragraph describes the methodology, and then the vulnerability analysis is presented for the five considered cases of study. Finally, some conclusions are drawn.

2. The Methodology

The five introduced indexes, which were evaluated in each node of the system, differ regarding the kind of information they can provide, and the way that they are formulated. The first considered parameter is the maximum TTHMs concentration value reached in each node during the entire simulation:

$$[TTHMs]_i^{max} = max\left\{ [TTHMs]_{i,t} \right\} \tag{1}$$

where $[TTHMs]_{i,t}$ is the actual TTHMs concentration at time t in the node i.

The second one is the TTHMs average concentration C_i^{TTHMs} weighted on nodes demand, which is defined as:

$$C_i^{TTHMs} = \frac{\int_0^{TOT} q_i(t) \cdot [TTHMs]_i \cdot dt}{\int_0^{TOT} q_i(t) \cdot dt} \tag{2}$$

where TOT is the total observation time, $q_i(t)$ is the actual demand node (L/s), and i represents the node index.

Two other useful indexes are the dimensionless exposure time T_i^E and the normalized contaminated volume $V_i^{c,n}$, which are expressed by:

$$T_i^E = \frac{\int_0^{TOT} I(t) \cdot dt}{TOT} \tag{3}$$

$$V_i^{c,n} = \frac{\int_0^{TOT} q_i(t) \cdot I(t) \cdot dt}{\int_0^{TOT} q_i(t) \cdot dt} \tag{4}$$

where the factor $I(t)$ is defined as:

$$I(t) = \begin{cases} 1 & if \quad [TTHMs]_{i,t} > [TTHMs]_{lim} \\ 0 & if \quad [TTHMs]_{i,t} \leq [TTHMs]_{lim} \end{cases} \tag{5}$$

$[TTHMs]_{lim}$ is a fixed attention threshold. T_i^E and $V_i^{c,n}$, represent the percentage of the time and the demand distributed in presence of contaminated water in a node during the observation period, respectively. They can assume values between 1, when all of the water that is delivered is contaminated, and 0, which means that the threshold is never exceeded in the node. Note that with fixed base demands in the nodes, there are no differences between the T_i^E and $V_i^{c,n}$, values. Moreover, in a node, both values are equal to 1 if the limit concentration is exceeded for the entire simulation period.

The last index is the contaminated mass load M_i^{TTHMs}, which corresponds to the mass of TTHMs delivered in each demand node when their concentration is above the threshold $[TTHMs]_{lim}$, and is formulated as:

$$M_i^{TTHMs} = \int_0^{TOT} \left([TTHMs]_{i,t} - [TTHMs]_{lim} \right) \cdot q_i(t) \cdot dt \tag{6}$$

As already mentioned, the five indexes furnish different and complementary information. In particular, both the maximum instantaneous and the average TTHMs concentrations are evidence of a likely dangerous exposure, without considering number of users involved and the duration of the exposition. The dimensionless exposure time and the normalized contaminated volume consider the duration of the exposition to dangerous values, while the contaminated mass is also related to the number of users exposed. The last three indexes indicate if there is a chronicle or sporadic TTHMs exposure above the alert bound.

For the evaluation of the indexes defined by Equations (1)–(6), the chlorine concentration is computed using the first-order kinetic equation, and assuming that it is consumed only through bulk reaction:

$$[Cl]_{i,t} = [Cl_2]_{i,0} \exp(-k_b \cdot t) \tag{7}$$

where $[Cl_2]_{i,0}$ (mg/L) and $[Cl_2]_{i,t}$ (mg/L) are the chlorine concentration initially and at time t for the node i, respectively; and k_b (1/h) is the bulk chlorine decay coefficient.

The TTHMs concentrations are computed as a linear function of consumed chlorine [37,42], using the following equation:

$$[TTHMs]_{i,t} = [TTHMs]_{i,0} + D \cdot \left([Cl_2]_{i,0} - [Cl_2]_{i,t} \right) \tag{8}$$

where $[TTHMs]_{i,0}$ (µg/L) and $[TTHMs]_{i,t}$ (µg/L) are the TTHMs concentrations initially and at time t in the node i, respectively; and D (µg/mg) is the TTHMs yield coefficient, which is defined as the ratio of µg of TTHMs formed to mg of chlorine consumed.

Both the chlorine decay and the TTHMs formation coefficients have to be properly assigned. Practically, they can be fixed considering literature suggestions or calibrated using measured data.

Hydraulic and water quality simulations are performed using the EPANET [43] and EPANET Multi-Species Extension (EPANET-MSX) codes. EPANET-MSX [44,45] enables the evaluation of chlorine and TTHMs concentrations through just providing the formulations governing the reaction dynamics. Using the EPANET and EPANET-MSX toolkits working simultaneously, the whole methodology is implemented in a code written in the C++ programming language.

3. Results

The proposed methodology is applied to five case studies (Figure 1): the example network Net-3 introduced in the EPANET User's Manual, the PW06 network by Prasad and Walter [46], the Anytown water distribution system [47], the main trunk of AVWSS (Aurunci-Valcanneto Water Supply System) located in Lazio region, Italy [36], and the real network of Cimitile, which is located in South Italy.

In all of the cases, water is supposed to be disinfected through sodium hypochlorite. For the reaction rate coefficients to model chlorine decay, k_b (Equation (7)) and TTHMs formation, D (Equation (8)), unique global values are assumed, because not significant variations of the environmental conditions have been registered in the networks. In particular, for the cases of Net-3, PW06, Anytown, and Cimitile, the following values are chosen from the literature: $k_b = 0.071 \times 1/h$ [48] and $D = 45$ µg/mg are included in the range identified by Bocelli et al. [49]. In the performed tests, sodium hypochlorite is supposed to be injected with a constant dosage equal to 0.68 mg/L, and the initial chlorine and TTHMs concentrations at any demand node are fixed equal to 0.68 mg/L and 10.1 µg/L, respectively. As regards AVWSS, the values of the rate coefficients for chlorine decay and TTHMs formation have been calibrated, which is explained in more detail in Section 3.4.

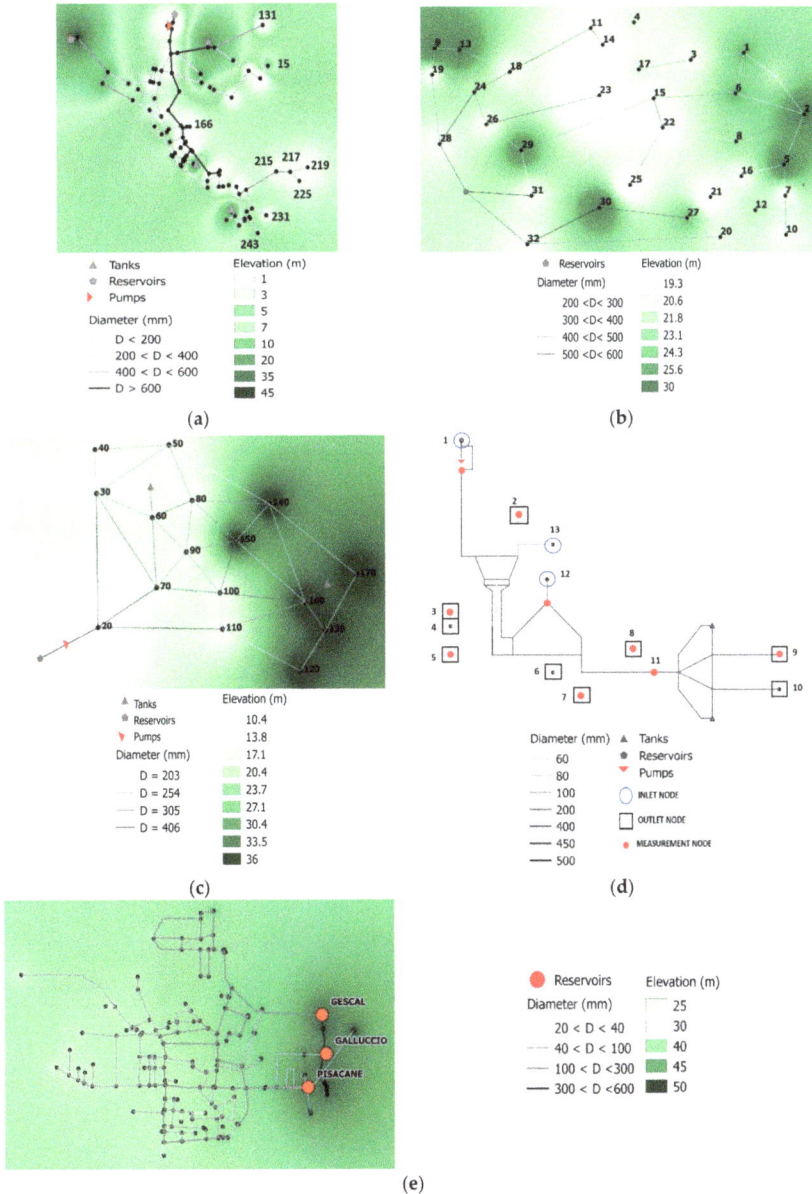

Figure 1. Scheme of case studies. (**a**) Net-3, (**b**) PW06, (**c**) Anytown, (**d**) Aurunci-Valcanneto Water Supply System (AVWSS), and (**e**) Cimitile.

Considering that the Italian regulatory limit for TTHMs is of 30 µg/L, the TTHMs concentration threshold is fixed equal to 25 µg/L as a precautionary measure. A different threshold can be adopted in a different country or when using the methodology for studying another DPB, through modifying the identified vulnerable areas and number of users exposed. For the hydraulic and water quality simulations, the time step is 5 min. The vulnerability indexes are computed considering 24 h of simulation ($TOT = 24$ h).

3.1. Net-3 Network

The literature case study Net-3 is composed of 117 pipes, 92 junctions, two pumps, and three tanks, and it is supplied by two sources that are modeled as reservoirs (Figure 1a). Geometric data, base demands, and patterns are reported in the EPANET User's Manual. The Hazen–Williams law is adopted as the resistance formula. Sodium hypochlorite is supposed to be injected in the two reservoirs.

Figure 2 presents the vulnerability contour plot maps obtained for the five considered indexes. The maps furnish complementary information about the different kind of exposure that they describe. Considering $[TTHMs]_i^{max}$ and C_i^{TTHMs} (Figure 2a,b), the green regions represent the "safe" zones, where the $[TTHMs]_{lim}$ is never exceeded. The vulnerable nodes identified with C_i^{TTHMs}, from yellow to red areas, are less than the ones identified considering $[TTHMs]_i^{max}$, because the latter parameter is more restrictive and considers the node as critical even if the imposed limit has been exceeded only at one time step. In particular, 59 nodes reach during the day a TTHMs concentration above the limit, while C_i^{TTHMs} values above the threshold are registered in only 18 nodes. In particular, the nodes 131, 166, and 243 have the maximum C_i^{TTHMs}, which is approximately equal to 37 µg/L. Moreover, the highest values of both $[TTHMs]_i^{max}$ and C_i^{TTHMs} are registered in node 243.

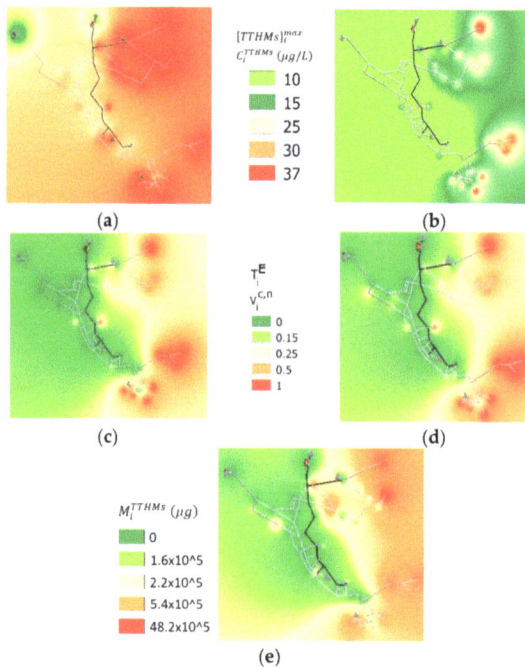

Figure 2. Vulnerability contour plot maps for Net-3: (**a**) maximum actual total trihalomethanes (TTHMs) concentration, (**b**) demand weighted average TTHMs concentration, (**c**) dimensionless exposure time, (**d**) normalized contaminated volume, and (**e**) contaminated mass load.

As expected, the maps obtained with $V_i^{c,n}$ and T_i^E (Figure 2c,d) are very similar, indicating that there are eight nodes (the red ones) in which the limit concentration is exceeded for all of the simulation period ($V_i^{c,n} = T_i^E = 1$). All of these nodes are characterized by high water age values. In this group, the node 243 is also included.

Finally, the computed M_i^{TTHMS} values range from 277 µg to about 485×10^4 µg, as reported in Figure 2e. The vulnerability map is slightly different from the other ones, and node 15, which is not included in the ones selected through $V_i^{c,n}$ and T_i^E, is individuated as the most vulnerable one. In fact, in this node, the water age is lower, but the base demand, representing the daily mean values of the outputs in the node, is higher, indicating a large number of users.

3.2. PW06 Network

The PW06 network (Figure 1b) has 47 pipes and 33 demand nodes with elevations that vary between 10–30 m, and it is supplied from a single source (reservoir), where sodium hypochlorite is supposed to be injected. The base demands are half of the ones reported in the original paper, and the demand pattern of Figure 3a has been assigned in all of the nodes. The Hazen–Williams law is adopted as the resistance formula.

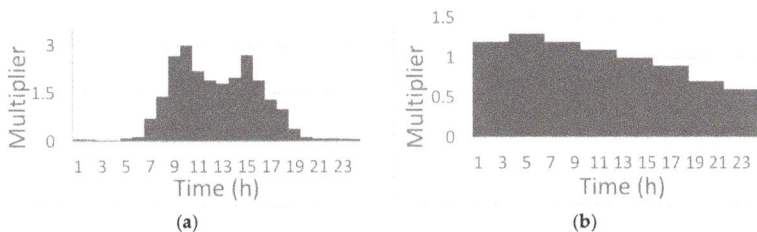

Figure 3. Demand patterns used in case studies. (**a**) PW06 and Cimitile, (**a,b**) Anytown.

Unlike the Net-3 system, in this case study, the $[TTHMs]_i^{max}$ and C_i^{TTHMs} values are very similar, as shown from the heat maps reported in Figure 4a,b. For both parameters, the limit is exceeded in the same 13 nodes, and they identify node 15 as the most vulnerable one.

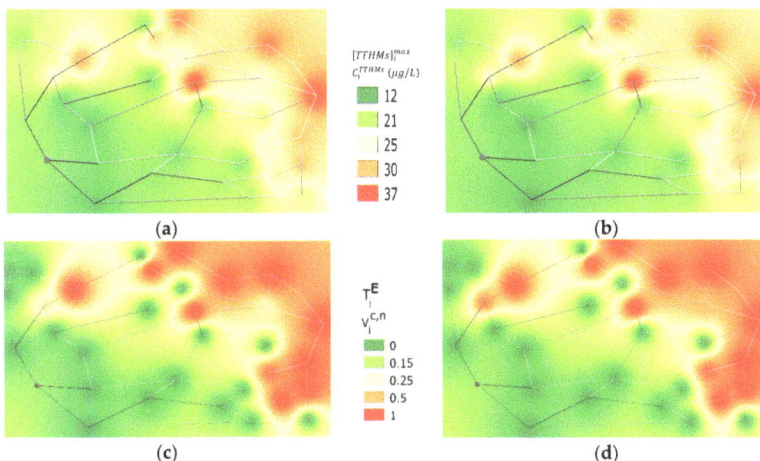

Figure 4. *Cont.*

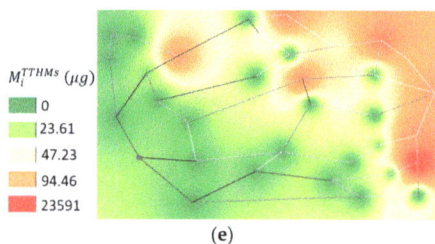

(e)

Figure 4. Vulnerability contour plot maps for PW06: (**a**) maximum actual TTHMs concentration, (**b**) demand weighted average TTHMs concentration, (**c**) dimensionless exposure time, (**d**) normalized contaminated volume, and (**e**) contaminated mass load.

As in the Net-3 case, the heat maps of T_i^E and $V_i^{c,n}$, as reported in Figure 4c,d, are similar, with 10 nodes in which the limit concentration is exceeded for the entire simulation period ($V_i^{c,n} = T_i^E = 1$). They are located in the opposite side with respect to the source, where the water ages have larger values. Also, the mass parameters individuate the same vulnerable area (Figure 4e), and the maximum M_i^{TTHMS} value ($M_i^{TTHMS} = 14,053 \times 10^3$ µg) is registered in node 7, which is included in the critical nodes selected through $V_i^{c,n}$ and T_i^E.

3.3. Anytown Network

The Anytown water distribution system (Figure 1c) has 16 junctions, two tanks, 34 pipes, three pumps, and one reservoir, where sodium hypochlorite is supposed to be injected. The network has been partially modified for the purposes of the paper. In particular, in the nodes 20, 50, 60, 80, and 130, the base demands have been reduced by one third with respect to the original values, and the pattern in Figure 3b has been assigned, while for the rest of the nodes, the base demands have been reduced by one fifth with respect to the original values, and the demand pattern of Figure 3a has been used. The Hazen–Williams law is adopted as the resistance formula for the simulation.

Figure 5 depicts the heat maps obtained with the five considered parameters. In this case study, unlike PW06, $[TTHMs]_i^{max}$ and C_i^{TTHMs} identify the same critical nodes, but the values that they assume are very different, as indicated from heat maps in Figure 5a,b. In 13 nodes, the TTHMs concentration exceeds the limit, while the C_i^{TTHMs} reaches the threshold in only six nodes. For both parameters, the most critical node is the number 170, which is located far from the source. Similarly to the Net-3 network, the comparison between Figures 5a and 5b indicates that the maximum concentration is a more restrictive parameter.

The parameters T_i^E and $V_i^{c,n}$ individuate the same nodes as critical, but their values are different, as indicated from the heat maps of Figure 5c,d. In particular, the values of the dimensionless exposure time are lower than the ones of the normalized contaminated volume. The limit concentration is exceeded for the entire simulation period ($V_i^{c,n} = T_i^E = 1$) in only nodes 140 and 170.

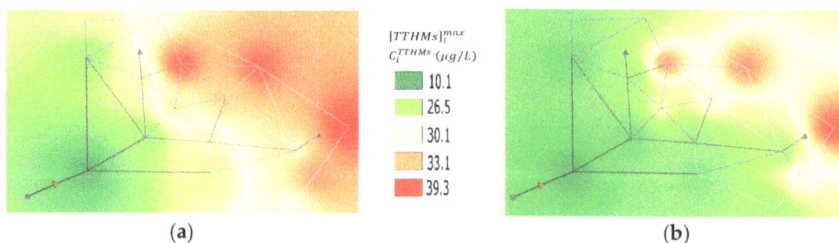

(a)   (b)

Figure 5. *Cont.*

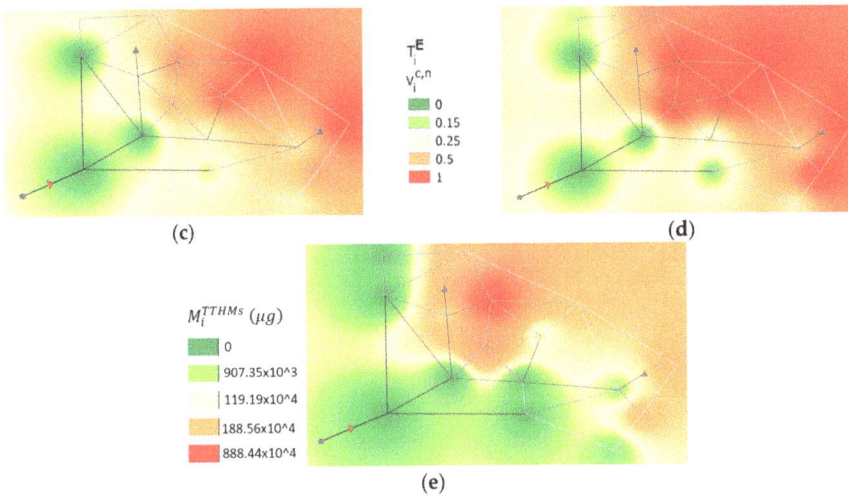

Figure 5. Vulnerability contour plot maps for Anytown: (**a**) maximum actual TTHMs concentration, (**b**) demand weighted average TTHMs concentration, (**c**) dimensionless exposure time, (**d**) normalized contaminated volume, and (**e**) contaminated mass load.

The heat map corresponding to the parameter M_i^{TTHMS} reported in Figure 5e is slightly different from the other ones, and the maximum value ($M_i^{TTHMS} = 41{,}756 \times 10^3$ µg) is observed at node 80, which is not included in the ones individuated from the other parameters.

3.4. Aurunci Valcanneto Water Supply System (AVWSS)

The considered main trunk of the AVWSS is located in the Lazio region, and it serves about 110,000 inhabitants distributed among 25 towns. As shown in Figure 1d, the system is supplied by three sources (inlet nodes 1, 12, and 13), and water is delivered in nine outlet nodes, which represent the insertion into nine different WDSs.

In November 2008, a campaign of water quality measurements was realized, and the measured values of chlorine and TTHMs concentrations ($[Cl_2]_m$ and $[TTHMs]_m$) are reported in Table 1. A better description of the AVWSS can be found in Di Cristo et al. [36], where the system has been used as a case study to compare and validate some empirical and kinetics models for the prediction of the TTHMs concentration.

The demands in the delivering nodes are reported in Table 1, and no pattern is assigned, since the AVWSS is a main trunk. The simulation is performed in steady condition, and the obtained water age in the nodes are reported in Table 1. The limit considered on TTHMs concentration is 1.5 µg/L, which is fixed considering the extension of the WDSs connected to the outlet nodes. In other words, this threshold corresponds to about the 25 µg/L at the more distant outputs nodes of the connected WDSs. The initial chlorine concentration is 0.30 mg/L, while the TTHMs one is null. The reaction rate coefficients to model chlorine decay, k_b (Equation (7)), and TTHMs formation, D (Equation (8)), are fixed through a specific calibration equal to 0.02 1/h and 43 µg/mL, respectively.

Without a pattern in the nodes, the parameter C_i^{TTHMs} does not make sense, and T_i^E and $V_i^{c,n}$ are coincident. As shown in Table 1, the results of the vulnerability analysis identify different critical nodes according to the different indexes. In particular, $[TTHMs]_i^{max}$ selects node 9 as the most vulnerable one (with a maximum value of 3.91 µg/L), which corresponds to the node with the maximum water age. T_i^E assume the value 0 in four nodes where the $[TTHMs]_{lim}$ has never been exceeded, and 1 in the nodes 5, 7, 8, 9, 10, corresponding to ages higher than 6 h. In terms of mass, the maximum value is delivered in node 10, with the maximum base demand ($M_i{}^{TTHMs}$ = 16,460.44 × 10^3 µg), which is included in the group identified by $T_i{}^E$.

Table 1. Water quality measurements of November 2008 and the results of the vulnerability analysis.

Node Number	Age (h)	Q (L/s)	$[Cl_2]_m$ (mg/L)	$[TTHMs]_m$ (µg/L)	$[TTHMs]_i{}^{max}$ (µg/L)	$T_i{}^E$	$M_i{}^{TTHMs}$ (10^3 µg)
2	0.17	2	0.27	1.97	0.04	0	0
3	1.89	8	0.27	1.49	0.46	0	0
4	3.92	2	-	-	0.94	0	0
5	6.85	2	0.26	1.32	1.65	1	26.28
6	5.98	8	-	-	1.45	0	0
7	6.37	2.5	-	2.00	1.54	1	9.33
8	8.57	3	-	2.45	2.03	1	138.36
9	18.06	47	0.24	3.02	3.91	1	9823.64
10	8.2	112	0.25	-	3.20	1	16,460.44

3.5. Cimitile Network

Figure 1e reports the scheme of the water distribution network of Cimitile, a small town located in the Campania region in Southern Italy. The system, which serves about 3700 residents located in 2.74 km^2, consists of 433 pipes, with a total length of approximately 20 km, five regulation valves, and 404 junction nodes. Pipe diameters range from 20 mm to 260 mm. Predominant pipe materials are cast iron and steel, while there is only a small percentage of high-density polyethylene (HDPE) pipes. The water comes from a regional main trunk (not reported in the scheme) and it goes into the system through three inlet nodes (Gescal, Galluccio, and Pisacane).

Base demands values, as shown in the scheme of Figure 6 have been obtained working out utility consumption billing data, and the daily total flow rate supplied to the users is about 34.25 m^3/s. The same water demand pattern, as characterized by two main peaks (Figure 3a), has been assigned to all of the nodes. The Darcy–Weisbach head loss formula is used for the simulation, and the hydraulic parameters have been previously calibrated using tank levels and pressure measurements. The chlorination of the water occurs in the three inlet nodes.

The vulnerability contour plot maps, as reported in Figure 7, indicate that also in this case, the vulnerable indexes tend to identify different risk areas considering different kinds of exposure. For a better understanding, Figure 6 reports the scheme of the network with the more vulnerable nodes selected using the different indexes. Moreover, Figure 8 illustrates the temporal variability of the TTHMs concentration in nodes 211, 290, and 328.

In terms of $[TTHMs]_i{}^{max}$ and $C_i{}^{TTHMs}$ (Figure 7a,b), the most critical nodes are located in the west part of the network. The number of nodes characterized by the excess of the attention limit in terms of maximum actual TTHMs concentration are more copious, because, as already mentioned, this parameter leads to also classifying as vulnerable those nodes where the $[TTHMs]_{lim}$ is exceeded only once during the simulation time. Both parameters individuate the node 211 as the most critical one (Figure 6), in which the maximum $C_i{}^{TTHMs}$ is about 27 µg/L. Furthermore, the same node 211 is located in the area of the network that is characterized by the highest water ages.

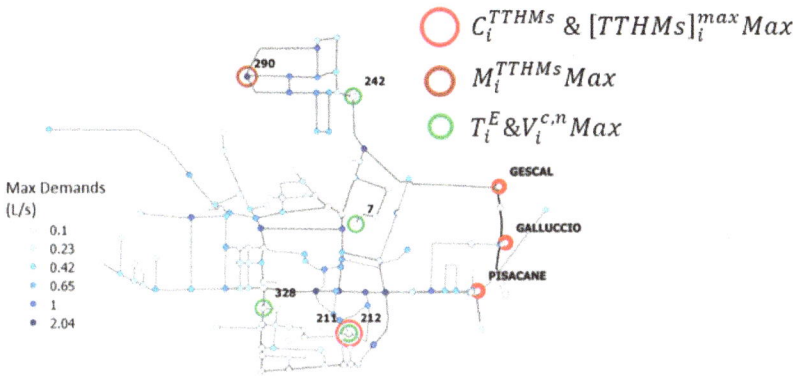

Figure 6. Scheme of the Cimitile network with the more critical nodes.

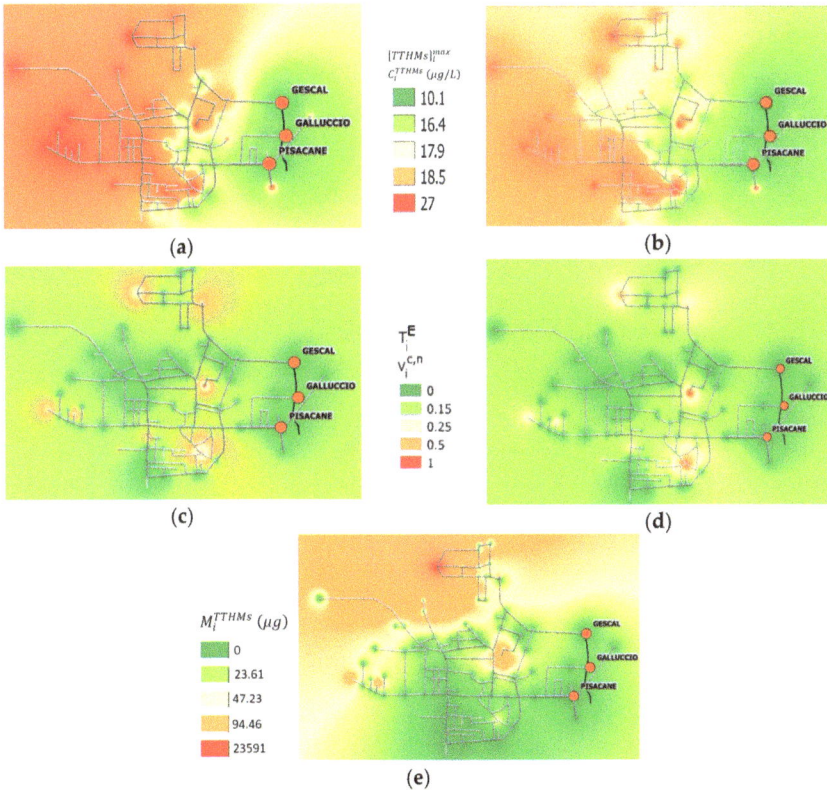

Figure 7. Vulnerability contour plot maps for Cimitile: (**a**) maximum actual TTHMs concentration, (**b**) demand weighted average TTHMs concentration, (**c**) dimensionless exposure time, (**d**) normalized contaminated volume, and (**e**), contaminated mass load.

The heat maps show that the exposure time and the normalized contaminated volume (Figure 7c,d) enable identifying the same critical areas in the system. In particular, the zones with high T_i^E and $V_i^{c,n}$ values are located in the north, central, and south parts of the network, and the lower and higher residence times nodes are equally involved. In five nodes 7, 211, 212, 242, and 328 (Figure 6), the concentration exceeds the limit for the entire simulation period ($V_i^{c,n} = T_i^E = 1$).

The M_i^{TTHMs} values range from 600×10^3 μg at node 214 to about 23×10^{12} μg at node 290 (Figure 7e). The high values of this parameter coincide with nodes characterized by large base demands, namely with a high number of users served. Using this index, the most vulnerable is node number 290, which is located in northwest region of the network, and is not included in the group individuated by the other parameters (Figure 6).

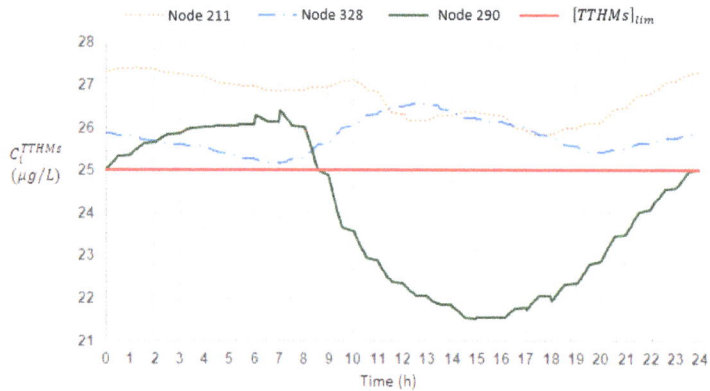

Figure 8. Temporal variation of the TTHMs concentration in nodes 211, 290, and 328.

Figure 8 reports the TTHMs concentration time variability during the simulating day for the nodes 211, 290, and 328, identified as the most vulnerable during the analysis.

In node 290, which is identified as the most critical one according to M_i^{TTHMs}, the TTHMs concentration exceeds the attention limit for about 8 h at the beginning of the simulation and in the last few minutes. In the other considered nodes, 211 and 328, which are vulnerable according to all of the parameters except M_i^{TTHMs}, the TTHMs concentration are always above the threshold fixed at 25 μg/L, reaching a maximum value of around 27 μg/L in node 211. These results indicate that in the area of node 290, many users were exposed to a high concentration for a short period, while in the areas of nodes 211 and 328, fewer consumers were always exposed to TTHMs concentrations above the fixed threshold. This kind of analysis is very useful for optimizing the network management for better water quality.

4. Conclusions

Water distribution vulnerability to elevated TTHMs concentrations is explored through the use of five different indexes that enable evaluating the population's exposure. The methodology is applied to five case studies: three literature schemes—namely Net-3, PW06, and Anytown—and two real systems, namely AVWSS and Cimitile, which are both located in Italy.

The main outcome of the obtained results is that the introduced indexes identify different critical areas with respect to elevated TTHMs concentrations, depending on the kind of exposure to which they refer. They enable identifying the higher risk nodes with respect to the different kinds of exposure: high concentration exposure for a short period, or chronic exposure to low TTHMs concentrations. The comparison between the maps obtained from the different indexes suggests that their performances strongly depend on the behavior of the system.

The presented research is significant because vulnerability analyses are important for water companies and administrators to design public health and strategies to reduce risks. The proposed methodology provides valuable information for assisting water utilities in the WDSs management considering water quality. In particular, the results of the presented study and the information embedded are also effective in the implementation of methodologies for the formulation of optimization problems such as chlorine dosage or chlorine booster allocation in the system or for planning network modifications. The vulnerability indexes can also be adopted for performing the district sectorization and skeletonization of a network accounting for water quality. Moreover, even if it is presented considering the TTHMs, the proposed methodology can be used for evaluating the exposure to others DPBs, adopting the adequate modelling for predicting the concentration and selecting the relative threshold limit.

Author Contributions: Conceptualization, all authors; Methodology, C.Q. and A.L.; Software, C.Q.; Validation, A.L., Formal Analysis, C.Q. and C.D.C. Data Curation, C.Q. and A.L.; Writing-Original Draft Preparation, C.Q. and C.D.C.; Writing-Review & Editing, C.Q. and C.D.C.

Funding:: This research received no external funding.

Conflicts of Interest: The authors declare no conflict of interest.

References

1. Vasconcelos, J.J.; Rossman, L.A.; Grayman, W.M.; Boulas, P.F.; Clark, R.M. Kinetics of chlorine decay. *J. Am. Water Works Assoc.* **1997**, *89*, 54–65. [CrossRef]
2. Gang, D.C.; Clevenger, T.E.; Banerji, S.K. Modeling Chlorine Decay in Surface Water. *J. Environ. Inform.* **2003**, *1*, 21–27. [CrossRef]
3. Greenberg, A.E. Public Health Aspects of Alternative Water Disinfectants. *J. Am. Water Works Assoc.* **1981**, *73*, 31–33. [CrossRef]
4. White, G.C. Ozone. In *Handbook of Chlorination Alternative Disinfectants*; Van Nostrand Reinhold: New York, NY, USA, 1992; pp. 1046–1110.
5. Morris, R.D.; Audet, A.M.; Angelillo, I.F. Chlorination, chlorination by-products and cancer: A metanalysis. *Am. J. Public Health* **1992**, *2*, 955–963. [CrossRef]
6. Krewski, D.; Balbus, J.; Butler-Jones, D.; Haas, C.; Isaac-Renton, J.; Roberts, K.J.; Sinclair, M. Managing Health Risks from Drinking Water—A Report to the Walkerton Inquiry. *J. Toxicol. Environ. Health Part A* **2002**, *65*, 1635–1823. [CrossRef] [PubMed]
7. Weinberg, H.S.; Krasner, S.W.; Richardson, S.W.; Thurston, A.D. *The Occurrence of Disinfection By-Product (DBPs) of Health Concern in Drinking Water: Results of a Nationwide DBP Occurrence Study*; EPA International Report 2002, EPA/600/R-02/068; EPA: Washington, DC, USA, 2002.
8. Waller, K.; Swan, S.H.; DeLorenze, G.; Hopkins, B. Trihalomethanes in Drinking Water and Spontaneous Abortion. *Epidemiology* **1998**, *9*, 134–140. [CrossRef] [PubMed]
9. Dodds, L.; King, W.D. Relation between Trihalomethane Compounds and Birth Defects. *Occup. Environ. Med.* **2001**, *58*, 443–446. [CrossRef] [PubMed]
10. Richardson, S.D.; Simmons, J.E.; Rice, G. Disinfection Byproducts: The next Generation. *Environ. Sci. Technol.* **2002**, *36*, 198–205. [CrossRef]
11. Chowdhury, S.; Champagne, P. Risk from exposure to trihalomethanes during shower: Probabilistic assessment and control. *Sci. Total. Environ.* **2009**, *407*, 1570–1578. [CrossRef] [PubMed]
12. Villanueva, C.M.; Cantor, K.P.; Grimalt, J.O.; Castaño-Vynials Malats, N.; Silverman, D.; Tardon, A.; Garcia-Closas, R.; Serra, C.; Carrato, A.; Rothman, N.; et al. Bladder cancer and exposure to disinfection byproducts in water through ingestion, bathing, showering and swimming in pools: Findings from the Spanish Bladder Cancer Study. *Am. J. Epidemiol.* **2007**, *165*, 148–156. [CrossRef] [PubMed]
13. Murray, R.; Janke, R.; Uber, J. The Threat Ensemble Vulnerability Assessment (TEVA) Program for Drinking Water Distribution System Security. In Proceedings of the World Water and Environmental Resources Congress, Salt Lake City, UT, USA, 27 June–1 July 2004.

14. Gunnarsdottir, M.J.; Gardarsson, S.M.; Elliott, M.; Sigmundsdottir, G.; Bartram, J. Benefits of Water Safety Plans: Microbiology, Compliance, and Public Health. *Environ. Sci. Technol.* **2012**, *46*, 7782–7789. [CrossRef] [PubMed]

15. Propato, M.; Piller, O.; Uber, J.G. A Sensor Location Model to Detect Contaminations in Water Distribution Networks. *Impacts Glob. Clim. Chang.* **2005**, 1–12. [CrossRef]

16. Janke, R.; Murray, R.; Uber, J.; Taxon, T. Comparison of Physical Sampling and Real-Time Monitoring Strategies for Designing a Contamination Warning System in a Drinking Water Distribution System. *J. Water Resour. Plan. Manag.* **2006**, *132*, 310–313. [CrossRef]

17. Maslia, M.L.; Sautner, J.B.; Aral, M.M.; Reyes, J.J.; Abraham, J.E.; Williams, R.C. Using Water-Distribution System Modeling to Assist Epidemiologic Investigations. *J. Water Resour. Plann. Manag.* **2000**, *126*, 180–198. [CrossRef]

18. Perelman, L.; Maslia, M.L.; Ostfeld, A.; Sautner, J.B. Using Aggregation/Skeletonization Network Models for Water Quality Simulations in Epidemiologic Studies. *J. Am. Water Works Assoc.* **2008**, *100*, 122–133. [CrossRef]

19. Saldarriaga, J.G.; García, S.; León, N. A Methodology to Preserve Water Quality Modeling in Skeletonized Network Models. In Proceedings of the World Environmental and Water Resources Congress, Albuquerque, NM, USA, 20–24 May 2012; pp. 2932–2942.

20. Weisel, C.P.; Kim, H.; Haltmeier, P.; Klotz, J.B. Exposure estimates to disinfection byproducts of chlorinated drinking water. *Environ. Health Perspect.* **1999**, *107*, 3–10. [CrossRef]

21. Li, X.-F.; Mitch, W.A. Drinking water disinfection byproducts (DBPs) and human health effects: Multidisciplinary challenges and opportunities. *Environ. Sci. Technol.* **2018**, *52*, 1681–1689. [CrossRef] [PubMed]

22. Krasner, S.W.; Cantor, K.P.; Weyer, P.J.; Hildesheim, M.; Amy, G. Case study approach to modeling historical disinfection by-product exposure in Iowa drinking waters. *J. Environ. Sci.* **2017**, *58*, 183–190. [CrossRef] [PubMed]

23. World Health Organization. *Guidelines for Drinking-Water Quality*; World Health Organization: Geneva, Switzerland, 2004; Volume 1.

24. Health Canada, Guidelines for Canadian Drinking Water Quality. 2007. Available online: http://www.hc-sc.gc.ca/ewhsemt/alt_formats/hecs-sesc/pdf/pubs/water-eau/2010-sum_guide-res_recom/sum_guide-res_recom-eng.pdf (accessed on 9 July 2018).

25. *National Health and Medical Research Council, Australian Drinking Water Guidelines*; 2004; Volume 6. Available online: http://www.nhmrc.gov.au/guidelines/publications/eh34 (accessed on 9 July 2018).

26. Propato, M.; Uber, J.G. Vulnerability of Water Distribution Systems to Pathogen Intrusion: How Effective Is a Disinfectant Residual? *Environ. Sci. Technol.* **2004**, *38*, 3713–3722. [CrossRef] [PubMed]

27. Murray, R.; Uber, J.; Janke, R. Model for Estimating Acute Health Impacts from Consumption of Contaminated Drinking Water. *J. Water Resour. Plan. Manag.* **2006**, *132*, 293–299. [CrossRef]

28. Nilsson, K.A.; Buchberger, S.G.; Clark, R.M. Simulating Exposures to Deliberate Intrusions into Water Distribution Systems. *J. Water Resour. Plan. Manag.* **2005**, *131*, 228–236. [CrossRef]

29. Khanal, N.; Buchberger, S.G.; McKenna, S.A. Distribution system contamination events: Exposure, influence, and sensitivity. *J. Water Resour. Plann. Manag.* **2006**, *132*, 283–292. [CrossRef]

30. Davis, M.J.; Janke, R. Importance of Exposure Model in Estimating Impacts When a Water Distribution System Is Contaminated. *J. Water Resour. Plan. Manag.* **2008**, *134*, 449–456. [CrossRef]

31. Davis, M.J.; Janke, R. Development of a Probabilistic Timing Model for the Ingestion of Tap Water. *J. Water Resour. Plan. Manag.* **2009**, *135*, 397–405. [CrossRef]

32. Thompson, S.L.; Casman, E.; Fischbeck, P.; Small, M.J.; VanBriesen, J.M. Vulnerability Assessment of a Drinking Water Distribution System: Implications for Public Water Utilities. In Proceedings of the World Environmental and Water Resources Congress, Tampa, FL, USA, 15–19 May 2007.

33. Baoyu, Z.; Xinhua, Z.; Yuan, Z. Vulnerability Assessment of Regional Water Distribution Systems. In Proceedings of the 2009 International Conference on Environmental Science and Information Application Technology, Wuhan, China, 4–5 July 2009; pp. 473–477.

34. Islam, N.; Sadiq, R.; Rodriguez, M.J.; Legay, C. Assessing Regulatory Violations of Disinfection By-Products in Water Distribution Networks Using a Non-Compliance Potential Index. *Environ. Monit. Assess.* **2016**, *188*, 304. [CrossRef] [PubMed]

35. Islam, N.; Sadiq, R.; Rodriguez, M.J. Optimizing Locations for Chlorine Booster Stations in Small Water Distribution Networks. *J. Water Resour. Plan. Manag.* **2017**, *143*, 04017021. [CrossRef]

36. Di Cristo, C.; Esposito, G.; Leopardi, A. Modelling trihalomethanes formation in water supply systems. *Environ. Technol.* **2013**, *34*, 61–70. [CrossRef] [PubMed]

37. Kavanaugh, M.C.; Trussell, A.R.; Cromer, J.; Trussell, R.R. An Empirical Kinetic Model of Trihalomethane Formation: Applications to Meet the Proposed THM Standard. *J. Am. Water Works Assoc.* **1980**, *72*, 578–582. [CrossRef]

38. Brown, D.; Bridgeman, J.; West, J.R. Understanding data requirement for trihalomethane formation in water supply systems. *Urban Water J.* **2011**, *8*, 41–56. [CrossRef]

39. Adin, A.; Katzhendler, J.; Alkaslassy, D.; Rav-Acha, C. Trihalomethane Formation in Chlorinated Drinking Water: A Kinetic Model. *Water Res.* **1991**, *25*, 797–805. [CrossRef]

40. Di Cristo, C.; Leopardi, A.; de Marinis, G. Assessing measurement uncertainty on trihalomethane prediction through kinetic models in water supply systems. *J. Water Supply Res. Technol. AQUA* **2015**, *64*, 516–528. [CrossRef]

41. Di Cristo, C.; Leopardi, A.; Quintiliani, C.; de Marinis, G. Drinking Water Vulnerability Assessment after Disinfection through Chlorine. *Procedia Eng.* **2015**, *119*, 389–397. [CrossRef]

42. Gang, D.C.; Segar, R.L.; Clevenger, T.E.; Banerji, S.K. Using Chlorine Demand to Predict TTHM and HAA9 Formation. *J. Am. Water Works Assoc.* **2002**, *94*, 76–86. [CrossRef]

43. Rossman, L. *EPANET User's Manual*; Risk Reduction Engineering Laboratory, U.S. Environmental Protection Agency: Cincinnati, OH, USA, 1994.

44. Shang, F.; Uber, J.G.; Rossman, L.A. Modeling Reaction and Transport of Multiple Species in Water Distribution Systems. *Environ. Sci. Technol.* **2007**, *42*, 808–814. [CrossRef]

45. Shang, F.; Uber, J.G.; Rossman, L.A. *EPANET Multi-Species Extension User's Manual*; Risk Reduction Engineering Laboratory, U.S. Environmental Protection Agency: Cincinnati, OH, USA, 2008.

46. Prasad, T.D.; Walters, G.A. Minimizing residence times by rerouting flows to improve water quality in distribution networks. *Eng. Optim.* **2006**, *38*, 923–939. [CrossRef]

47. Walski, T.M.; Brill, E.D.; Gessler, J.; Goulter, I.C.; Jeppson, R.M.; Lansey, K.; Lee, H.; Liebman, J.C.; Mays, L.; Morgan, D.R.; et al. Battle of networks models: Epilogue. *J. Water Resour. Plan. Manag.* **1987**, *113*, 191–203. [CrossRef]

48. Elshorbagy, W.A. Kinetics of THM Species in Finished Drinking Water. *J. Water Resour. Plan. Manag.* **2000**, *126*, 21–28. [CrossRef]

49. Boccelli, D.; Tryby, M.E.; Uber, J.G.; Summers, R.S. A reactive species model for chlorine decay and THM formation under rechlorination conditions. *Water Res.* **2003**, *37*, 2654–2666. [CrossRef]

water | MDPI

Article

Developing a Statistical Model to Improve Drinking Water Quality for Water Distribution System by Minimizing Heavy Metal Releases

Wei Peng and Rene V. Mayorga *

Faculty of Engineering, University of Regina, Regina, SK S4S 0A2, Canada; Wei.Peng@uregina.ca
* Correspondence: Rene.Mayorga@uregina.ca; Tel.: + 1-(306)-585-4397; Fax: + 1-(306)-585-4855

Received: 14 May 2018; Accepted: 6 July 2018; Published: 14 July 2018

Abstract: This paper proposes a novel statistical approach for blending source waters in a public water distribution system to improve water quality (WQ) by minimizing the release of heavy metals (HMR). Normally, introducing a new source changes the original balanced environment and causes adverse effects on the WQ in a water distribution system. One harmful consequence of blending source water is the release of heavy metals, including lead, copper and iron. Most HMR studies focus on the forecasting of unfavorable effects using precise and complicated nonlinear equations. This paper uses a statistical multiple objectives optimization, namely Multiple Source Waters Blending Optimization (MSWBO), to find optimal blending ratios of source waters for minimizing three HMRs in a water supply system. In this paper, three response surface equations are applied to describe the reaction kinetics of HMR, and three dual response surface equations are used to track the standard deviations of the three response surface equations. A weighted sum method is performed for the multi-objective optimization problem to minimize three HMRs simultaneously. Finally, the experimental data of a pilot distribution system is used in the proposed statistical approach to demonstrate the model's applicability, computational efficiency, and robustness.

Keywords: water quality (WQ); blending; release of heavy metals (HMR); dual response surface optimization (DRSO); multiple source waters blending optimization (MSWBO)

1. Introduction

Mixing water from a new source may alter the original balanced environment in water networks. As a consequence, pipe corrosion, decreasing disinfectant residual and microbiological growth will deteriorate WQ. Optimizing the blending ratio of water coming from different sources may minimize this water deterioration [1–4]. However, an optimal blending ratio usually cannot be realized because an operational blending ratio is dependent on the water demand, capacities of water plates, and locations of water plates and the mixing point of source waters. Furthermore, the concentrations of heavy metals are unassignable by water plates. They are impacted by the chemical parameter concentrations (such as chlorine concentration) and physical parameter levels (such as turbidity), also influenced by the original balanced environment in water networks. It is a challenge to minimize the release of heavy metals (HMR) in a water distribution system.

Imran et al. [1,5] described a method that optimize the blending of multiple source waters for heavy metal corrosion abatement and monochloramine residual control in distribution systems. They built up a pilot water distribution system to simulate a full-scale operational water distribution system. They fitted the nonlinear empirical WQ models for copper, lead, apparent color, and monochloramine dissipation. In addition, then based on the regulations, the optimal water blending ratio ranges were found. However, this optimization may need to be further developed, to provide some quantitative indicators to water utilities for improving the WQ in their water distribution systems. This paper proposes a new statistical optimization method for offering such quantitative indicators.

The optimization of this study is nonlinear because of the nonlinearity of the WQ. Conventional nonlinear optimization methodologies, such as gradient-based algorithms [6–8] and genetic algorithms [9], have some disadvantages that may not be suitable for use in the field of WQ. E.g., a gradient-based algorithm may have a large development cost and the linearization can be time-consuming; the genetic algorithms are time-consuming when be used in a large-scale nonlinear problem. Here, in this paper a second-order statistical model, namely the dual response surface optimization (DRSO), is introduced to conduct this nonlinear optimization.

The DRSO is based on the Response Surface Method (RSM). The RSM is a statistical technique used in empirical study. It approximates the true response surface and estimates the parameters; it works well for solving real response surface problems [10,11]. As a consequence, it searches for an optimal set of input variables to optimize the response by using a set of designed experiments. In the past few years, several robust DRSO techniques have been developed [12,13]. Yanikoglu et al. (2016) introduced Taguchi's Robust Parameter Design approach into DRSO to develop a method that uses only experimental data, and it can yield a solution that is robust against ambiguity in the probability of inputs [12].

The optimization is also required to run rapidly because it faces the dynamics of WQ, water demand, and source water capacity of water plates [14,15]. RSM programming can be quickly solved by any commercial solver because it contains a quadratic model rather than a high-order nonlinear model. All observations in RSM are typically assumed to have equal variation that may not be practically valid [16]. Vining and Myers [17] then proposed a DRSO model that includes two quadratic empirical models, one for the mean and another for the standard deviation, and then optimized one of the responses subjected to an appropriate constraint given by the other. Because the second empirical model considers the standard deviation, a DRSO approach can avoid misleading optimum results and produce robust results.

Harmful chemical reactions (especially for HMR) and microbiological growth are required to be controlled in water distribution systems. According to the previous studies, the biomass accumulation is influenced to a greater extent by the nature of the supporting material (such as unlined ductile iron) than by the secondary parameters of WQ [18]. Meanwhile, Imran et al. [1,5] confirmed that loss of disinfectant residual is affected to a greater extent by the delivery distance and retention time rather than the WQ. This paper only focuses on the relationships between the HMR and the chemical and physical parameters of WQ in the distribution system. Three heavy metals are iron, lead, and copper.

This paper proposes a Multiple Source Waters blending Optimization (MSWBO) methodology to find optimal blending ratios for multiple source waters according to the WQ parameters of these source waters (shown in Table 1) and a variety of scenario designs. The objectives of the study are to reduce three HMR of lead, copper, and iron in water distribution systems, and minimize the operational cost. A weighed sum method is used for the multi-objective function in the MSWBO to minimize HMR simultaneously. The dual response surface models are used to describe the reaction kinetics of three HMR, with tracking the standard deviations of the responses of these three release equations. An application is provided to demonstrate the applicability and advantages of this MSWBO.

Table 1. Water quality parameters of the source waters.

Parameter	GW	SW	DW
Alkalinity (mg/L) as $CaCO_3$	225	50	50
Calcium (mg/L) as $CaCO_3$	200	50	50
Dissolved oxygen (mg/L)	8	8	8
pH	7.9	8.2	8.3
Silica (mg/L)	14	7	1

Table 1. *Cont.*

Parameter	GW	SW	DW
HRT (day)	5	5	5
Sodium (mg/L)	10	15	30
Chloride (mg/L)	15	10	50
Sulfates (mg/L)	10	180	30
Temperature (°C)	25	25	25

Notes: Sourced from Imran et al. [1]; GW is groundwater, SW is surface water; DW is desalination water. HRT is hydraulic retention time.

2. Methodology

The development of the proposed MSWBO approach consists of three main steps: data acquisition, model fitting and optimization. Data acquisition includes data type identification, data range determination, experiment design, and experimental data recording.

2.1. Data Acquisition

2.1.1. Data Type Identification and Range Determination

Three HMR are fitted into the dual response surface models. For the dual response surface models, three mean responses are the release levels (RLs) of iron, lead, and copper in the mixed water; three standard deviation responses are the standard deviations of the responses of these three release equations; the input variables are the concentrations of the parameters of WQ in the source waters, respectively. These WQ parameters include alkalinity, calcium, dissolved oxygen, pH, silica, sodium chloride, sulfates, hydraulic retention time (HRT), and temperature.

For the MSWBO, the responses of the dual response surface models are treated as "the control variables"; the operational cost of the WQ adjustment of source waters is also treated as "the control variable"; the input variables of the dual response surface models are treated as "the state variables" or "the decision variables", which will be discussed in the next section; because the ratios of groundwater (GW), surface water (SW), and desalination water (DW) in the blending are given by the decision maker, they are treated as the state variables.

The value ranges of the variables related to WQ are limited by government standards such as National Primary Drinking Water Regulations (Primary Standards) and National Secondary Drinking Water Regulations (Secondary Standards) [19,20]. The sum of three ratios is equal to 1.

2.1.2. Experiment Design and Experimental Data Recording

Tampa Bay Water (TBW) operates a water distribution system and was required by the Southwest Florida Water Management District to reduce groundwater withdrawal for minimizing hydrological and ecological impacts to the riverine systems and ensuring that flows remain within the range of natural variability [21]. To complete this requirement, TBW completed the constructions of a surface water plant and a desalination plant [1,5]. Meanwhile, a study [1] was sponsored by TBW and the American Water Works Association (AWWA) Research Foundation to evaluate the WQ impacts from the new source water supplies. As a part of the above-mentioned study, a pilot distribution system (PDS) was constructed in Florida, U.S., to simulate the full-scale operational water distribution system of TBW [1,5]. Three source waters were selected in the PDS: a conventional groundwater (GW), surface water (SW) and desalinated water (DW). Table 1 describes their WQ information. The aged pipes of the PDS were removed from the water distribution system of TBW. The pipes were assembled on the pilot site and allowed to equilibrate with groundwater over a period of five months. After equilibrium was established, various blends were introduced into the PDS. Two years of continuous operation was performed on the PDS which was used to study the relationship between WQ and the distribution system.

2.2. Model Fitting

Box and Wilson [22] first used the Response Surface Method to study the relationship between a response and a set of input variables. Vining and Myers [11] first fitted second-order polynomial models for a mean and standard deviation separately and optimized one of the responses subjected to an appropriate constraint given by the other. Since the second empirical model considers the standard deviation, a MSWBO approach can avoid misleading optimum and produce robust results. A general dual response surface model is formulated as follows:

$$C_\mu = h_0 + S'h + S'HS \tag{1}$$

$$C_\sigma = g_0 + S'g + S'GS \tag{2}$$

where $g_0 = \alpha_0, h_0 = \beta_0, g = (\alpha_1, \alpha_2, \ldots \alpha_k)', h = (\beta_1, \beta_2, \ldots \beta_k)'$, and

$$H = \frac{1}{2} \begin{bmatrix} 2\beta_{11} & \beta_{12} & \cdots & \beta_{1k} \\ \beta_{21} & 2\beta_{22} & \cdots & \\ \vdots & \vdots & \ddots & \vdots \\ \beta_{k1} & \beta_{k2} & \cdots & 2\beta_{kk} \end{bmatrix} \tag{3}$$

$$G = \frac{1}{2} \begin{bmatrix} 2\alpha_{11} & \alpha_{12} & \cdots & \alpha_{1k} \\ \alpha_{21} & 2\alpha_{22} & \cdots & \\ \vdots & \vdots & \ddots & \vdots \\ \alpha_{k1} & \alpha_{k2} & \cdots & 2\alpha_{kk} \end{bmatrix} \tag{4}$$

where h_0, h, g_0, g, H and G are the appropriate scalars, $(k \times 1)$ vectors, and $(k \times k)$ matrices for the estimated coefficients; C_μ and C_σ are the mean and standard deviation; S and S' are $(k \times 1)$ vectors of the input variables and their transpose, respectively.

The proposed study needs to fit three couples of second-order polynomial models: (1) the lead release model and its standard deviation model; (2) the copper release model and its standard deviation model; and (3) the iron release model and its standard deviation model. It should be noticed that these three couples of models use same secondary WQ parameters (alkalinity, sodium, pH, conductivity, etc.) as input variables.

2.3. Dual Response Surface Optimization

The next step is to optimize the above response surface models simultaneously. The objectives represent the desire for minimizing (1) metal RLs; (2) standard deviations of metal RLs; (3) adjustment costs of WQ for the source waters. With the constraints related to the above control variables, a MSWBO model is formulated as follows:

$$\text{Min} \quad f = \sum_{j=1}^{n} \{ w_1 (C_\mu)_j^2 + w_2 (C_\sigma)_j^2 + w_3 [x_{ki}(U_{ki} - \hat{U}_{ki})]_j^2 \} \tag{5a}$$

s.t.

$$(C_\mu)_j = (h_o + S'h + S'HS)_j \le (L_\mu)_j \tag{5b}$$

$$(C_\sigma)_j = (g_0 + S'g + S'GS)_j \tag{5c}$$

$$S_k = \sum_{i=1}^{m} x_i U_{ki} \tag{5d}$$

$$\underline{U_{ki}} \le U_{ki} \le \overline{U_{ki}} \tag{5e}$$

$$0 \leq x_i \leq 1 \tag{5f}$$

$$\sum_{i=1}^{m} x_i = 1 \tag{5g}$$

where j is the index of the response of metal release, $j = 1, 2, \ldots, n$; i is the index of the source water, $i = 1, 2, \ldots, m$; L_μ is the maximum permissible metal RL; S_k is the concentration of the kth input variable in the blend, k is the index of input variables; x_i is the percentage of the source water i in the blend, i is the index of source waters; U_{ki} is the concentration of the kth input variable in the source water i; $\hat{U}_{ki}$ is the original concentration of the kth input variable in the source water i; $\underline{U_{ki}}$ and $\overline{U_{ki}}$ are the lower bound and upper bound of U_{ki}; w_1, w_2 and w_3 are the weights to reflect different priorities to the corrosion and the cost of WQ adjustment.

U_{ki} is the decision variable. It directly affects the WQ of the blend. Meanwhile, the lower the difference between U_{ki} and $\hat{U}_{ki}$, the low the operation cost of water utilities. In this study, C_μ, C_σ and $(U_{ki} - \hat{U}_{ki})$ are included in the objective function to reflect the trade-off between operation cost and WQ. This can be realized by adjusting values of w_1, w_2 and w_3 as shown in Equation (5a).

3. Application

3.1. Nonlinear Empirical Model

The previous studies formulated the statistical nonlinear corrosion models for copper, lead and color (iron) [1,5,23,24]. The nonlinear release models are listed as follows:

$$\text{Cu} = (\text{T})^{0.72}(\text{Alk})^{0.73}(\text{pH})^{-2.86}(\text{SO}_4)^{0.1}(\text{SiO}_2)^{-0.22} \tag{6}$$

$$\text{Pb} = 1.027^{(\text{T}-25)}(\text{Alk})^{0.677}(\text{pH})^{-2.726}(\text{Cl})^{1.462}(\text{SO}_4)^{-0.228} \tag{7}$$

$$\Delta\text{C} = 10^{-1.321}(\text{T})^{0.813}(\text{Alk})^{-0.912}(\text{Cl})^{0.485}(\text{Na})^{0.561}(\text{SO}_4)^{0.118}(\text{DO})^{0.967}(\text{HRT})^{0.836} \tag{8}$$

where Cu and Pb are the copper and lead concentrations in mg/L, ΔC is the increase in apparent color (measured in cpu: Co-Pt unit), T is temperature in °C, Alk is the concentration of alkalinity in mg/L as calcium carbonated (CaCO_3); SO_4 and SiO_2 are the concentrations of sulfates and silica in mg/L, respectively, Cl is the concentration of chlorides in mg/L, pH is the dimensionless measure of the acidity or alkalinity of a solution, and (HRT) is the hydraulic retention time in days.

Imran et al. (2006) concluded a strong relationship existed between the total iron (Fe) concentration (mg/L) and the apparent color (cpu) [1]:

$$Fe = 0.0132 \times Apparent\ Color \tag{9}$$

$$R^2 = 0.82$$

where R^2 is the correlation coefficient.

The major advantage of the above nonlinear models is explicitly describing the relationships between the metal RLs and the concentrations of secondary WQ parameters, but these nonlinear models will cause the optimization to be time consuming.

3.2. Dual Response Surface Model

The dual response surface model is composed of two second-order polynomial models, one for a mean and another for the standard deviation. The mean model of a metal release describes the relationship between the metal RLs and the concentrations of WQ parameters. The standard deviation model is to establish the relationship between the standard deviation and the WQ parameters. If it only uses the mean model for optimization, the result may have a large variation. The second empirical model of metal release considers the standard deviation; thus, the optimization can optimize the metal release subjected to its standard deviation constraints. The dual response surface optimization can avoid generating such a large variation result and then can produce a robust result.

The WQ parameter of pH is non-conservative. It is defined as a negative decimal logarithm of the molar concentration of the hydrogen ion activity in a solution. Because the pH value of blends cannot be obtained from a mass balance equation, we chose the molar concentration of the activity of hydrogen ions A_{H+} as a substitution ($A_{H+} = 10^{-pH}$).

The next step is to convert the above metal nonlinear models into the dual response surface models. The corresponding dual response surface models are formulated as follows:

Iron-release response surface model

$$
\begin{aligned}
C_\mu^{Fe} =\ & 10^{-3}(-9763 + 355T - 2Alk + 17Cl - 43Na - 6.5SO_4 + 1259DO + 699HRT \\
& - 0.1T\bullet Alk - 0.4T\bullet Cl + 1T\bullet Na + 0.1T\bullet SO_4 - 21T\bullet DO + 3.9T\bullet HRT \\
& + 0.1Alk\bullet Na - 0.3Alk\bullet DO - 0.2Alk\bullet HRT + 0.1Cl\bullet Na - Cl\bullet DO \\
& + 0.9Cl\bullet HRT + 1.9Na\bullet DO + 0.2Na\bullet HRT + 0.5SO_4\bullet DO \\
& + 0.2SO_4\bullet HRT + 8.2DO\bullet HRT \\
& - 4.2T^2 + 0.03Alk^2 - 0.04Cl^2 + 0.1Na^2 - 47.7DO^2 - 118.4HRT^2) \\
& R^2 = 0.817
\end{aligned}
\tag{10}
$$

$$
\begin{aligned}
C_\sigma^{Fe} =\ & 10^{-4}(-65091 - 123T + 23Alk - 73Cl + 196Na - 39SO_4 + 2601DO + 37656HRT \\
& - 0.3T\bullet Alk + 1.7T\bullet Cl - 4T\bullet Na + 0.7T\bullet SO_4 - 20T\bullet DO + 11\,T\bullet HRT \\
& - 0.1Alk\bullet Na - 1.4Alk\bullet DO + 0.1Alk\bullet HRT \\
& - 0.1Cl\bullet Na + 5.2Cl\bullet DO + 0.9Cl\bullet HRT \\
& - 13.8Na\bullet DO - 0.1Na\bullet HRT \\
& + 2.7SO4\bullet DO + 0.2SO_4\bullet HRT + 0.3DO\bullet HRT \\
& + 6.9T^2 - 0.01Alk^2 - 0.04Cl^2 + 0.28Na^2 - 0.01SO_4{}^2 - 112.7DO^2 - 5431HRT^2) \\
& R^2 = 0.990
\end{aligned}
\tag{11}
$$

where C_μ^{Fe} is the mean response of iron release in mg/L, C_σ^{Fe} is the standard deviation response of iron release; T is the temperature in °C; Alk is the concentration of alkalinity in mg/L as calcium carbonate ($CaCO_3$); Na, SO_4 and Cl are the concentrations of sodium, sulfates and chlorides in mg/L, respectively; DO is the dissolved oxygen content in mg/L; and HRT is the hydraulic retention time in days. R^2 is the correlation coefficient; it should be noticed that the value of R^2 in Equation (10) is calculated by the quadratic equation responses against the results of Equation (8).

Copper-release response surface model

$$
\begin{aligned}
C_\mu^{Cu} =\ & 10^{-3}(-252.3 + 35T + 2.4Alk - 12.3A_{H+} + 1.4SO_4 - 15SiO_2 \\
& + 0.1T\bullet Alk + 1.2T\bullet A_{H+} - 0.5T\bullet SiO_2 \\
& + 0.04Alk\bullet A_{H+} + 0.01Alk\bullet SO_4 - 0.14Alk\bullet SiO_2 \\
& + 0.04A_{H+}\bullet SO_4 + 0.34A_{H+}\bullet SiO2 - 0.08SO_4\bullet SiO_2 \\
& - 0.77T^2 - 0.68A_{H+}{}^2 + 1.34SiO_2{}^2) \\
& R^2 = 0.997
\end{aligned}
\tag{12}
$$

$$C_\sigma^{Cu} =$$
$$10^{-4}(-1132 + 124T + 4.8Alk - 77A_{H+} + 3SO_4 - 6SiO_2$$
$$- 0.2T\bullet Alk + 2.6T\bullet A_{H+} - 0.1T\bullet SO_4 + 0.3T\bullet SiO_2$$
$$+ 0.09Alk\bullet A_{H+} - 0.03Alk\bullet SiO_2 \tag{13}$$
$$+ 0.14A_{H+}\bullet SO_4 + 1.27A_{H+}\bullet SiO_2 - 0.17SO_4\bullet SiO_2$$
$$- 2.92T^2 - 0.01Alk^2 - 0.81A_{H+}{}^2 - 0.07SiO_2{}^2)$$
$$R^2 = 0.999$$

where C_μ^{Cu} is the mean response of copper release in mg/L, C_σ^{Cu} is the standard deviation response of copper release; T is the temperature in °C; Alk is the concentration of alkalinity in mg/L as calcium carbonate (CaCO$_3$); SO$_4$ and SiO$_2$ are the concentrations of sulfates and silica in mg/L, respectively; A_{H+} is the molar concentration of the activity of hydrogen ions in 10^9 mol/L (in order to keep the same order of magnitude as the pH value). R^2 is the correlation coefficient.

Lead-release response surface model

$$C_\mu^{Pb} =$$
$$10^{-3}(6239 - 159T + 49Alk - 1527A_{H+} - 257Cl + 66SO_4$$
$$-0.2T\bullet Alk + 35.3T\bullet A_{H+} + 8.3T\bullet Cl - 2T\bullet SO_4$$
$$+ 3.76Alk\bullet A_{H+} + 1.25Alk\bullet Cl - 0.29Alk\bullet SO_4 \tag{14}$$
$$+ 11.6A_{H+}\bullet Cl + 1.8A_{H+}\bullet SO_4 - 1.4Cl\bullet SO_4$$
$$-1.2T^2 - 0.26Alk^2 - 18.6A_{H+}{}^2 + 2.18Cl^2 + 0.12SO_4{}^2)$$
$$R^2 = 0.993$$

$$C_\sigma^{Pb} =$$
$$10^{-4}(-49578 + 4041T + 188Alk - 2854A_{H+} + 471Cl + 50SO_4$$
$$-8.7T\bullet Alk + 62.3T\bullet A_{H+} - 5.8T\bullet Cl - 8T\bullet SO_4$$
$$+ 11Alk\bullet A_{H+} - 1.5Alk\bullet Cl + 0.004Alk\bullet SO_4 \tag{15}$$
$$-5.4A_{H+}\bullet Cl + 10.3A_{H+}\bullet SO_4 + 0.4Cl\bullet SO_4$$
$$-65.5T^2 - 0.22Alk^2 - 13.8A_{H+}{}^2 - 1.1Cl^2 + 0.05SO_4{}^2)$$
$$R^2 = 0.998$$

where C_μ^{Pb} is the mean response of lead release in mg/L, C_σ^{Pb} is the standard deviation response of lead release; T is the temperature in °C; Alk is the concentration of alkalinity in mg/L as calcium carbonate (CaCO$_3$); SO$_4$ and SiO$_2$ are the concentrations of sulfates and silica in mg/L, respectively; A_{H+} is the molar concentration of the activity of hydrogen ions in 10^9 mol/L (in order to keep the same order of magnitude as the pH value). R^2 is the correlation coefficient.

3.3. MSWBO Approach

3.3.1. Maximum Permissible Metal Release

The Lead and Copper Rule of US Environmental Protection Agency (EPA) for copper stipulates 90% of samples have a copper concentration less than 1.3 mg/L, and the concentration for lead is less than 15 µg/L [19]. The goal of this study is to find optimal blending ratios of source waters that can minimize the HMR in the distribution system. The minimized heavy metal concentrations should be less than the values of the regulations. Assuming these HMR followed a normal distribution; the maximum permissible metal RLs are their mean values. We know the standard deviations of the copper and lead RLs are 0.22 mg/L and 7 µg/L, respectively, and a one tailed Z statistic for 90% compliance is 1.28. The mean values of the copper and lead RLs are estimated on 1.02 mg/L and 6.04 µg/L, respectively. The actual government lead RL data indicated that the model (7) over-predicted the lead RL [1,5]. The maximum permissible lead release is then modified on a value of 10 µg/L.

The modification has been experimented with in the governments' distribution systems and shows a better reflection on the distribution function of lead release [1,5].

The EPA guides the secondary maximum contaminate level of iron is 0.3 mg/L [20]. We assume an influent iron RL of 0.1 mg/L, the maximum permissible iron RL within the distribution system is set at 0.2 mg/L. The metal release constraints are listed as:

$$L_{Cu} = 1 \ mg/L; \ L_{Pb} = 10 \ \mu g/L; \ L_{Fe} = 0.2 \ mg/L \tag{16}$$

where L_{Cu}, L_{Pb}, and L_{Fe} represent the maximum permissible concentrations of copper, lead, and iron in a water distribution system, respectively.

3.3.2. Variable Design

The task of this study is to find the optimal blending ratio of source waters, which consequently minimize all harmful reactions in the distribution system, as well as minimize the quality adjustment costs of source waters. The WQ chemical parameters of the source waters are considered as the decision variables. It is not necessary to use all of these chemical parameters as decision variables in the optimization model, such as some non-sensitive chemical parameters. Because these non-sensitive chemical parameters do not make much contribution to the metal release, e.g., the concentration of dissolved oxygen almost does not make contribution to the HMR, they are not necessary to be adjusted in the source waters. Hence, chemical parameters are considered as state inputs (certain value inputs). In practice, it is easy to increase the concentration of a chemical parameter in water rather than remove it from water, except for pH value. Therefore, the decision variables are designed to be varied from their initial value to a designed upper bound.

We found that only temperature, alkalinity and sulfates are common impacting parameters in the three metal release models (Equations (6)–(8)). These three common impacting parameters are selected as decision variables. Due to the parameter of temperature as an uncontrollable parameter which is dependent on ambient temperature; it is set on 25 °C in this study. Because the initial concentrations of alkalinity in the GW and sulfates in SW are high, they are considered as the state inputs.

The pH is a major importance in determining the corrosivity of water. In general, the higher pH value, the lower HMR. Equations (6) and (7) show the effects of increasing pH for copper and lead control. In water treatment process, addition of calcium hydroxide $Ca(OH)_2$ is a technology for water pH increasing. So that, in real-time operation increasing pH value of water is easier than reducing it. In this study, pH is replaced by A_{H+}, namely the molar concentration of the activity of hydrogen ions, due to the A_{H+} value in a blending source waters can be calculated by mass balance. The SMCL's for pH is 6.5 to 8.5, and the initial pH values for the GW, SW and DW are 7.9, 8.2 and 8.3, respectively. So that, we design the pH of GW is a decision variable, and the value range is from 7.9 to 8.5. The pH of the SW and DW are state inputs because they are high initial values.

Equation (6) also demonstrates that increasing concentration of silica contributes to lower copper release. Silica is then designed as a decision variable in the SW and DW rather than in the GW because of a high initial concentration of silica in the GW.

A higher chloride and sodium concentrations may be expected because of membrane deterioration in the desalination plant. Equations (7) and (8) display that a higher concentration of the chloride can cause a higher release of lead and iron, and a higher concentration of the sodium leads to a higher iron release. Because the primary objective of the MSWBO is to minimize releases of the metals, the optimization model will inhibit any increase in chloride and sodium concentrations. Chloride and sodium have to be treated as the state inputs. For the same reason, the parameters of HRT and dissolved oxygen are also considered as the state inputs (independent variables) to the model. Table 2 lists the initial values of the state inputs and the ranges of the decision variables.

Table 2. Value range of water quality parameter.

Parameter	GW	SW	DW
Alkalinity (mg/L) as $CaCO_3$	225	[50, 100]	[50, 100]
Dissolved oxygen (mg/L)	8	8	8
Silica (mg/L)	14	[7, 14]	[1, 14]
HRT (day)	5	5	5
Sodium (mg/L)	10	15	30
Chloride (mg/L)	15	10	50
Sulfates (mg/L)	[10, 80]	180	[30, 80]
pH	[7.9, 8.5]	8.2	8.3
Temperature (°C)	25	25	25

3.4. Results and Discussion

The dual response surface models are established for the releases of iron (Equations (10) and (11)), copper (Equations (12) and (13)) and lead (Equations (14) and (15)) in a particular water distribution system. On the basis of the ALs, the maximum copper and lead RLs are set at 1.0 mg/L and 10 µg/L, respectively. The maximum iron RL is established at 0.2 mg/L in accordance with the SMCLs. According to Model 5, the MSWBO approach for this particular water distribution system is developed when subjected to the dual response surface models and the relative constraints. The objectives are to minimize the metal corrosions of the pipes and minimize the adjustment costs of WQ for the multiple source waters. The decision variables of the MSWBO are selected from the WQ parameter, and the value ranges of these variables are listed in Table 2.

Three source waters are used to study the blending source waters problem. The WQ information of the three source waters are listed in Table 1. Theoretically, an optimal blending ratio may exist for a minimum release. A distribution system needs different blends of source waters, which depend on where the water demand is located on. It is not appropriate to set a constraint on source water percentages as long as there was no centralized point of blending. On the other hand, to study infinite blends of source waters is unrealistic. For brevity, only a few extreme situations and threshold blending combinations are discussed here.

The blending ratio design in the extreme situations considers the situation of single source water supply and one source off line; the threshold blending combinations are obtained from the conclusions of previous studies [1,5], such as GW contributions of >60% result in unacceptable copper release and <20% result in an unacceptable release in color. For the purpose of comparison, the release simulations for the base scenario (WQ information is shown in Table 1) are conducted though the three metal release models directly. The results are listed in Table 3.

In the objective function of the MSWBO, the term of the metal concentrations represents the WQ of blending source waters and the term of the changes of the variables represents the operation cost. The minimization of the objective reflects the trade-off between WQ and operation cost. This can be realized thought adjusting values of w_1, w_2 and w_3 as shown in Equation (5a). Here, two weighting sets are designed as w_1: w_2: w_3 = 1: 1: 1 and w_1: w_2: w_3 = 100: 100: 1.

Examination of the MSWBO indicates that the unacceptable copper release, because of the effect of high GW percentage, can be addressed by increasing the pH value and/or silica concentration in the GW. Even for 100% of the GW, the copper RL can meet its maximum permissible RL though adjusting the pH value from 7.9 to 8.3 (Tables 4 and 5). The results of the MSWBO show that high pH value could help in controlling copper and lead release.

Table 3. Results of release models.

x	y	z	Pb	Cu	Fe
1	0	0	3.53	**1.15**	0.08
0.8	0.2	0	2.34	**1.12**	0.01
0.6	0	0.4	8.13	0.96	0.03
0.43	0.41	0.16	2.91	0.97	0.01
0.2	0.8	0	2.74	0.88	0.04
0	0.1	0.9	6	0.66	**0.39**
0	1	0	4.2	0.76	**0.21**
0	0.9	0.1	3.26	0.75	**0.24**
0.2	0.6	0.2	2.54	0.86	0.1
0.1	0.4	0.5	3.28	0.76	**0.25**
0.7	0.2	0.1	3.48	**1.07**	0.01
0.1	0	0.9	8.34	0.7	**0.3**
0	0	1	7.64	0.65	0.4
0.19	0	0.81	8.75	0.75	**0.22**
0.1	0.3	0.6	4.11	0.75	**0.26**
0.5	0.1	0.4	6.71	0.94	0.03
0.7	0.1	0.2	5.27	**1.04**	0.01
0.1	0.9	0	3.39	0.83	0.11

Notes: x is the blending ratio of GW, y is the blending ratio of SW, z is the blending ratio of DW; Pb is in μg/L, others are in mg/L; Boldface number indicates the excess of maximum RL.

Table 4. Results of MSWBO with $w_1 = 1$, $w_2 = 1$ and $w_3 = 1$.

x	y	z	Pb	Cu	Fe	Alk SW	Alk DW	SiO$_2$ SW	SiO$_2$ DW	SO$_4$ GW	SO$_4$ DW	pH GW
1	0	0	3.05	1	0.08	-	-	-	-	10	-	8.3
0.8	0.2	0	1.94	1	0.01	100	-	14	-	10	-	8.35
0.6	0	0.4	7.72	0.89	0.03	-	50	-	14	20.3	36.9	7.95
0.43	0.41	0.16	2.99	0.89	0.01	50	50	7	1	10	30	7.93
0.2	0.8	0	2.75	0.86	0.04	50	-	8.56	-	10	-	7.9
0	0.1	0.9	9.4	0.93	**0.21**	100	100	7	1	-	30	-
0	1	0	4.2	0.72	0.2	51.1	-	8.7	-	-	-	-
0	0.9	0.1	3.33	0.75	0.2	58.5	50.9	8.7	1.19	-	30	-
0.2	0.6	0.2	2.54	0.84	0.1	50	50	8.24	1.41	10	30	7.9
0.1	0.4	0.5	3.8	0.8	0.2	62.1	65.2	7.91	2.14	10	30	7.9
0.7	0.2	0.1	3.01	1	0.01	100	100	7	1	10	30	8.45
0.1	0	0.9	10	0.89	0.2	-	89.9	-	3.4	11.8	46	7.9
0	0	1	10	0.7	**0.24**	-	100	-	10.4	-	53.8	-
0.19	0	0.81	9.26	0.75	0.2	-	58.9	-	2.65	10	30	7.9
0.1	0.3	0.6	5	0.81	0.2	61.9	73.7	7.71	2.42	10	30	7.9
0.5	0.1	0.4	6.71	0.93	0.03	50	50	7.22	1.89	10	30	7.9
0.7	0.1	0.2	5.37	0.99	0.01	64.2	78.4	14	14	10	30	7.95
0.1	0.9	0	3.39	0.79	0.11	50	-	8.64	-	10	-	7.9

Notes: x is the blending ratio of groundwater (GW), y is the blending ratio of surface water (SW), and z is the blending ratio of desalination water (DW); Pb is in μg/L, others are in mg/L; Boldface number indicates the excess of maximum RL.

Table 5. Results of MSWBO with $w_1 = 100$, $w_2 = 100$ and $w_3 = 1$.

x	y	z	Pb	Cu	Fe	Alk SW	Alk DW	SiO$_2$ SW	SiO$_2$ DW	SO$_4$ GW	SO$_4$ DW	pH GW
1	0	0	3.05	1	0.08	-	-	-	-	10	-	8.3
0.8	0.2	0	1.94	1	0.01	100	-	14	-	10	-	8.35
0.6	0	0.4	7.72	0.89	0.03	-	50	-	14	20.3	36.9	7.95
0.43	0.41	0.16	2.99	0.89	0.01	50	50	7	1	10	30	7.93
0.2	0.8	0	2.75	0.78	0.04	50	-	14	-	10	-	7.91
0	0.1	0.9	9.4	0.93	**0.21**	100	100	7	1	-	30	-
0	1	0	4.2	0.64	0.2	51.1	-	14	-	-	-	-
0	0.9	0.1	3.33	0.64	0.2	58.5	50.9	14	8.04	-	30	-
0.2	0.6	0.2	2.54	0.74	0.1	50	50	14	13.9	10	30	7.91
0.1	0.4	0.5	3.8	0.66	0.2	62.1	65.2	14	14	10	30	8.19
0.7	0.2	0.1	3.23	1	0.01	100	100	14	1	10	30	8.27
0.1	0	0.9	10	0.67	0.2	-	89.9	-	14	11.8	46	7.91
0	0	1	10	0.7	**0.24**	-	100	-	10.4	-	53.8	-
0.19	0	0.81	9.13	0.62	0.2	-	60.9	-	14	11.2	35.1	7.91
0.1	0.3	0.6	5	0.65	0.2	61.9	73.7	14	14	10	30	7.9
0.5	0.1	0.4	6.58	0.84	0.03	50	50	14	14	14.6	33.7	7.92
0.7	0.1	0.2	5.37	0.99	0.01	64.2	78.4	14	14	10	30	7.95
0.1	0.9	0	3.39	0.71	0.11	50	-	14	-	10	-	7.9

Notes: x is the blending ratio of groundwater (GW), y is the blending ratio of surface water (SW), and z is the blending ratio of desalination water (DW); Pb is in μg/L, others are in mg/L; Boldface number indicates the excess of maximum RL.

The results of the base scenario simulations (Table 3) demonstrate that the distribution system does not have a heavy metal release problem when it receives the SW that has a percentage over the 90% in the blend waters, and it is only mixed with GW. If the SW were mixed with DW and the percentage of DW excessed 10% in the blend waters; then, high iron release will occur. The test of the MSWBO shows that the SW can satisfy the original balance environment of the pipe system in any percentage, by slightly increasing the concentration of silica and/or alkalinity. For instance, 100% of SW can be distributed to the pipe system with adjusting the concentration of silica from 7 mg/L to 8.7 mg/L and the concentration of alkalinity from 50 mg/L to 51.1 mg/L (Table 4).

High DW percentages (>90%) will result in unacceptable iron release, no matter how the WQ parameters are adjusted. However, this unacceptable iron release is much lower than that of the base scenario. For example, 100% of DW would cause 0.24 mg/L of iron release after optimally adjusting its alkalinity concentration (Table 4); for the base scenario, 100% of DW would result in 0.4 mg/L of iron RL (Table 3).

Examination of the MSWBO also verifies a previous observation: "corrosion of copper and lead pipes is increased by increasing alkalinity, whereas increasing alkalinity is beneficial in reducing the release of iron products from pipes; increasing sulfates reduces lead release but increases iron release" [1–3]. These conflicting WQ requirements for release abatement can be addressed by the mathematical optimization technology of MSWBO. The trade-off between WQ and operation cost can be realized by adjusting the weighting values of w_1, w_2 and w_3. Tables 4 and 5 list the results of the MSWBO with different weighting sets of w_1: w_2: w_3 = 1: 1: 1 and w_1: w_2: w_3 = 100: 100: 1, respectively. The first weighting set shows that decision makers pay more attention to operation cost; the latter represents that decision makers are focused on WQ.

It is worth mentioning that the development of the MSWBO is based on two assumptions: (1) the pipes are of fixed geometry and the flow is in low conditions; (2) the changes of the WQ parameters (such as alkalinity and sulfates) have no effect (or have a very slight effect) on the pH value. The future studies could be: (a) develop a full-scale calibration of the model for the application of higher flow rates; (b) develop a rigorous approach to evaluating the pH of bends.

The MSWBO to the metal release problem assumes the means and variances of these metals are known. It is found that the optimal solutions are quite sensitive to these assumptions. We can use historical data to estimate the means and variances. However, the noises are rarely known and that would result in a non-optimal solution [25]. These unknown noises need to be considered in a future study.

4. Conclusions

The Multiple Source Waters Blending Optimization (MSWBO) has been developed for blending source waters to improve WQ in a distribution system by minimizing the HMR. The developed MSWBO approach has favorable advantages over other approaches in the current literature; in particular it has the following properties:

- It provides a quantitative optimal blending ratio of source waters to water utilities for minimizing the HMR in their water distribution systems.
- It has a wide range of applications, because the experiment of this model was designed in a variety of scenarios, even for some extreme situations.
- It shows a high computational efficiency, due to fact that the model only includes some second-order equations and inequalities, and also that the experiment showed that the run time is only several milliseconds.
- It exhibits a robust operation, since the MSWBO model considers the standard deviation and avoids ambiguity in the probability of inputs.

Author Contributions: Conceptualization, W.P. and R.V.M.; Methodology, W.P.; Formal Analysis, W.P. and R.V.M.; Investigation, W.P.; Writing-Original Draft Preparation, W.P.; Writing-Review & Editing, W.P. and R.V.M.; Supervision, R.V.M.; Funding Acquisition, R.V.M.

Funding: This paper research has been supported by a grant (No: 155147-2013) from the Natural Sciences and Engineering Research Council of Canada (NSERC).

Acknowledgments: The authors would like to thank Syed Imran for providing some data.

Conflicts of Interest: The authors declare no conflict of interest.

References

1. Imran, S.; Dietz, J.D.; Mutoti, G.; Xiao, W.; Taylor, J.S.; Desai, V. Optimizing Source Water Blends for Corrosion and Residual Control in Distribution Systems. *J. AWWA* **2006**, *98*, 107–115. [CrossRef]
2. Kanakoudis, V.; Tsitifli, S. Potable Water Security Assessment: A Review on Monitoring, Modelling and Optimization Techniques, Applied to Water Distribution Networks. *Desalin. Water Treat.* **2017**, *99*, 18–26. [CrossRef]
3. Peng, W.; Mayorga, R.V.; Imran, S. A rapid fuzzy optimization approach to source water blending problems in water distribution systems. *Urban Water J.* **2012**, *9*, 177–187. [CrossRef]
4. Peng, W.; Mayorga, R.V.; Imran, S. A robust optimization approach for real-time multiple source water blending problem. *J. Water Supply Res. Technol.* **2012**, *61*, 111–122. [CrossRef]
5. Imran, S.; Dietz, J.D.; Mutoti, G.; Taylor, J.S.; Randall, A.A.; Cooper, C.D. Red Water Release in Drinking Water Distribution Systems. *J. AWWA* **2005**, *97*, 93–100. [CrossRef]
6. Mehrez, C.; Percia, C.; Oron, G. Optimal Operation of a Multisource and Multi-quality Regional Water System. *Water Resour. Res.* **1992**, *128*, 1199–1206. [CrossRef]
7. Ostfeld, A.; Shamir, U. Design of Optimal Reliable Multi-Quality Water Supply Systems. *J. Water Resour. Plan. Manag. ASCE* **1996**, *122*, 322–333. [CrossRef]
8. Yang, S.L.; Sun, Y.-H.; Yeh, W.W.-G. Optimization of Regional Water Distribution System with Blend Requirements. *J. Water Resour. Plan. Manag. ASCE* **2000**, *26*, 229–235. [CrossRef]
9. Tu, M.Y.; Tsai, F.T.-C.; Yeh, W.W.-G. Optimization of Water Distribution and Water Quality by Hybrid Genetic Algorithm. *J. Water Resour. Plan. Manag. ASCE* **2005**, *131*, 431–440. [CrossRef]

10. Anderson-Cook, C.M.; Borror, C.M.; Montgomery, D.C. Response surface design evaluation and comparison. *J. Stat. Plan. Inference* **2009**, *139*, 629–641. [CrossRef]

11. Myers, R.H.; Montgomery, D.C.; Vining, G.G.; Borror, C.M.; Kowalski, S.M. Response surface methodology: A retrospective and literature survey. *J. Qual. Technol.* **2004**, *36*, 53–77. [CrossRef]

12. Yanikoglu, I.; den Denhertog, D.; Kleijnen, J.P.C. Robust Dual-Response Optimization. *IIE Trans.* **2016**, *48*, 298–312. [CrossRef]

13. Huang, T.L.; Song, X.W.; Liu, X.Y. The Multi-objective Robust Optimization of The Loading Path in the T-shape Tube Hydroforming Based on Dual Response Surface Model. *Int. J. Adv. Manuf. Technol.* **2016**, *82*, 9–12. [CrossRef]

14. Benayoun, R.; De Montgolfier, J.; Tergny, J.; Laritchev, O. Linear programming with multiple objective functions: STEP method (STEM). *Math. Program.* **1971**, *1*, 366–375. [CrossRef]

15. Kuhn, H.W.; Tucker, A.W. Nonlinear programming. In *Mathematical Statistics and Probabilities*; Neyman, J., Ed.; Berkeley Symphony: Berkeley, CA, USA, 1951.

16. Lin, D.; Tu, W. Dual response surface optimization. *J. Qual. Technol.* **1995**, *27*, 34–39. [CrossRef]

17. Vining, G.G.; Myers, R.H. Combining taguchi and response surface philosophies: A dual response approach. *J. Qual. Technol.* **1990**, *22*, 38–45. [CrossRef]

18. Chang, Y.C.; Jung, K. Effect of Distribution System Materials and Water Quality on Heterotrophic Plate Counts and Biofilm Proliferation. *J. Microbial. Biotechnol.* **2004**, *14*, 1114–1119.

19. The Lead and Copper Rule. Available online: https://www.epa.gov/dwreginfo/lead-and-copper-rule (accessed on 11 July 2018).

20. The Secondary Maximum Contaminant Levels. Available online: https://www.epa.gov/dwstandardsregulations/secondary-drinking-water-standards-guidance-nuisance-chemicals (accessed on 11 July 2018).

21. Tampa Bay Water (TBW). Available online: http://www.tampabaywater.org (accessed on 11 July 2018).

22. Box, G.E.P.; Wilson, K.B. On the Experimental Attainment of Optimum Conditions. *J. R. Stat. Soc.* **1951**, *B*, 1–45.

23. Taylor, J.; Dietz, J.D.; Randall, A.A.; Hong, S.K.; Norris, C.D.; Mulford, L.A.; Arevalo, J.M.; Imran, S.; Le Puil, M.; Liu, S.; et al. *Effects of Blend on Distribution System Water Quality*; AwwaRF, Repor No. 91065F; Denver; Water Environment Research Foundation: Alexandria, VA, USA, 2005.

24. Xiao, W. Effect of Source Water Blend on Copper Release in a Pipe Distribution System: Thermodynamic and Empirical Models. Ph.D. Thesis, University of Central Florida, Orlando, FL, USA, 2004.

25. Bingham, D.; Nair, V.N. Noise variable settings in robust design experiments. *Technometrics* **2012**, *54*, 388–397. [CrossRef]

water

MDPI

Article

Estimation of Non-Revenue Water Ratio Using MRA and ANN in Water Distribution Networks

Dongwoo Jang and Gyewoon Choi *

Department of Civil & Environmental Engineering, Incheon National University, Incheon 22012, Korea;
nightray@paran.com
* Correspondence: gyewoon@inu.ac.kr

Received: 19 November 2017; Accepted: 18 December 2017; Published: 21 December 2017

Abstract: The non-revenue water (NRW) ratio in water distribution networks is the ratio of losses from unbilled authorized consumption and apparent and real losses to the total water supply. NRW is an important parameter for prioritizing the improvement of a water distribution system and identifying the influencing parameters. Though the method using multiple regression analysis (MRA) is a statistical analysis method for estimating the NRW ratio using the main parameters of a water distribution system, it has disadvantages in that the accuracy is low compared to the measured NRW ratio. In this study, an artificial neural network (ANN) was applied to estimate the NRW ratio to improve assessment accuracy and suggest an efficient methodology to identify related parameters of the NRW ratio. When using an ANN with the optimal number of neurons, the accuracy of estimation was higher than that of conventional statistical methods, as with MRA.

Keywords: non-revenue water; multiple regression analysis; artificial neural network; water distribution network

1. Introduction

Civil infrastructure such as bridges, buildings, and pipelines ensure economic and industrial prosperity in a society. Specifically, water distribution systems assure the delivery of the primary commodity of water [1]. They are, however, subject to deterioration over time that usually leads to problems such as the reduced utility of hydraulic facilities, water loss, service disruption, and lower water quality. Additionally, the gradual rise in consumer demand for water creates new problems such as low pressure at demand junctions. This usually raises the entrance water pressure of the water distribution system, which in turn increases the frequency of leakage [2].

To overcome problems in pressure management and ensure continuous, efficient, and economic operation of a water distribution system, an effective rehabilitation strategy is required. The latter should consider criteria on hydraulics, economics, reliability, and water quality performance, and [3] since the economic resources available for the rehabilitation of a water distribution system are scarce, it should also help prioritize investment [4].

In the early 1990s, no standard term existed to express and assess water losses in a water distribution system. The International Water Association (IWA) has acknowledged this problem and established a water loss task force (WLTF). The WLTF examined international best practices and developed a standardized terminology for non-revenue water (NRW) [5].

The IWA has recently proposed new performance indicators and their successful applications are being reported regularly [6,7]. A percentage indicator was also suggested not to be used for performance comparison, especially where target areas see large differences in consumption per service connection [8,9].

Analysis of the effects of pipe damage in water distribution networks determines the improvement priorities for each district water supply system; a systematic plan for replacement and remediation is required to maintain a city waterworks [10,11]. The waterworks improvement projects of old distribution networks are being implemented, but what is difficult is making management more economical and upgrading the operating system through the evaluation of aged pipes and the prevention of accidents based on empirical decisions [12].

Therefore, decision-making on the priority of maintenance of a water distribution network requires the study and analysis of the factors affecting leaks, as well as identifying the physical and operational parameters affecting leaks with parameters such as water pressure, and quality and demand quantity. To decrease the NRW ratio, studies on water distribution analysis, reliability enhancement, diagnosis of pipe network technology and pipe deterioration evaluation for optimal water distribution were conducted.

It is important to determine the level of contribution of leaks and bursts to the overall NRW volume [13] to evaluate the influence of water tariffs on NRW under a water supply plan [14]. Winarni (2009) found a performance indicator for facility management in water supply systems by comparing the infrastructure leakage index (ILI) [15]. Efforts to reduce leaks and commercial losses cost money, especially if large sections of piping need replacement. Nevertheless, studies have shown that efforts toward conservation and NRW reduction can provide water at about half to a third of the cost of water production [16]. A variety of studies have been carried out on the effective parameters of a water distribution network [6,17–19].

The index of the NRW ratio of a water distribution system needs to be proven by correlation with the region's characteristics and quantifying the influence of related parameters. For example, in areas with severe deterioration, the NRW ratio could be considered high because of many leaks, and it is difficult to find the factor of leakage parameters. In addition, unless the relationship between the regional characteristics and the NRW ratio is properly identified, the estimation of the ratio could prove unrealistic and uneconomic, even if the ratio is high due to local specificity.

In this study, a NRW ratio estimation model was suggested by using an artificial neural network (ANN) and multiple regression analysis (MRA). The statistical method was used to compare the results of ANN and MRA with the real measured values of the NRW ratio. This study also proposed a methodology for estimating the NRW ratio using ANN, which has the main parameters of water distribution system as independent variables and the NRW ratio as the dependent variable. To verify the suggested ANN and MRA, a target area was selected and applied.

2. General Research on the NRW Ratio and Theoretical Background

2.1. Non-Revenue Water of a Water Distribution Network

NRW corresponds to the percentage of water lost due to leaks and commercial problems, such as the lack of precision or mistakes in client databases. In Equation (1), A_p is the volume of water produced per time unit and A_b is the volume of billed water per time unit [2].

$$\text{NRW} = \frac{A_p - A_b}{A_p} \% \tag{1}$$

The definition of NRW could be described as follows. NRW is the difference between the volume of water put into a water distribution system and that billed to customers. The NRW ratio comprises three components [2].

(a) Physical losses comprise leaks from all parts of a water distribution system and overflow at water storage tanks. They can be caused by poor operations and maintenance, lack of active leakage control, and poor quality of underground assets.

(b) Commercial losses are caused by under-registration of customer meters, data handling errors, and theft of water in various forms.

(c) Unbilled authorized consumption includes water used by a utility for operational purposes, that used in firefighting, and that provided free to certain consumer groups.

Components of water balance in a water distribution system are shown in Table 1. One of the main elements of NRW is leakage, and finding and improving related factors in a district metered area (DMA) can lower the NRW ratio. In addition, it is important to identify the major factors influencing a water distribution network for NRW estimation.

Table 1. Components of Water Balance. Non-revenue water (NRW).

Water Produced	Effective water	Revenue water	Billed and metered
			Exported
			Others
		Effective NRW	Supplier's use
			Public use
			Illegal use
			Metering under-registration
	Ineffective water	Ineffective NRW	Discounted
			Leaks

2.2. Multiple Regression Analysis

Regression analysis is used to identify specific relationships between variables with high correlations or predict the value of variables. In the regression model, one independent variable is called simple regression analysis, and one independent variable does not sufficiently explain the dependent variable. When introducing multiple independent variables to the regression model, MRA is used with a linear function.

The multiple linear regression model with independent variables is expressed as Equation (2) [20].

$$y = \beta_0 + \beta_1 x_1 + \beta_2 x_2 + \ldots + \beta_k x_k \tag{2}$$

where x_i ($i = 1, \ldots, k$) is the independent variable, y is the dependent variable, β_i ($i = 1, \ldots, k$) is the regression coefficients, β_0 is the intercept of y, and β_0 is the slope of the independent variables.

2.3. Artificial Neural Network

The ANN procedure used is a feed-forward network type with input, hidden and output layers, as shown in Figure 1. Neurons in the input layer simply act as a buffer for distributing the input signals to neurons in the hidden layer. The neurons in different layers are interconnected via weights. The neurons in the hidden and output layers are called the activation function.

The activation function used here is a sigmoidal activation function. The input for each neuron j in the hidden layer is the sum of the weighted input signal x_i ($\sum_{i=1}^{n} w_{ji} x_i = net_j$, in which w_{ji} is the interconnecting weight between neuron j in the hidden layer and neuron i in the input layer). The output y_j from the neuron is given by Equation (3).

$$y_j = f\left(\sum_{i=1}^{n} w_{ji} x_i\right) = \frac{1}{\left(1 + e^{-net_j}\right)} \tag{3}$$

The output of neurons in the output layer is computed in a similar fashion. Schematic diagram of a multilayer feed-forward neural network is shown in Figure 1.

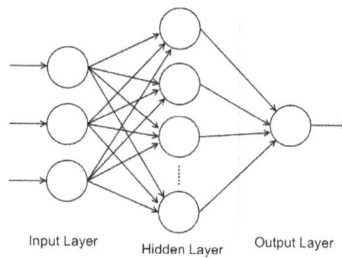

Figure 1. Schematic Diagram of Multilayer Feed-Forward Neural Network [21].

3. Methodology for Estimating the NRW Ratio Using MRA and ANN

The main parameters of the NRW ratio were analyzed and a classification system based on physical and operational parameters in water distribution systems was designated. Parameters that directly affect estimation of the NRW ratio are selected and the equation for estimating the ratio is introduced through MRA.

A parameter selection system was suggested through MRA, which is most widely used for predicting the NRW ratio [22,23]. Methods for evaluating accuracy are also described in detail per the entire process of estimating the NRW ratio, as shown in Figure 2.

When setting up the ANN model, effective parameters in water distribution networks are required. For the analysis using accuracy assessment of the ANN simulation results with the measured values, MRA is used to select statistically significant parameters and the NRW ratio was estimated by applying selected parameters to ANN. For this purpose, the selection of the target area and collection of data on the water distribution system are required. The influential factors selected are among the physical and operational parameters of the water distribution system; the water pressure in the pipe network can be considered through the demand energy ratio given by Jo [24].

The statistical procedure performs MRA with various physical and operational parameters as independent variables and the NRW ratio as a dependent variable. MRA is needed to select the most influential parameters of the ratio, and can also be used for estimating the ratio [25].

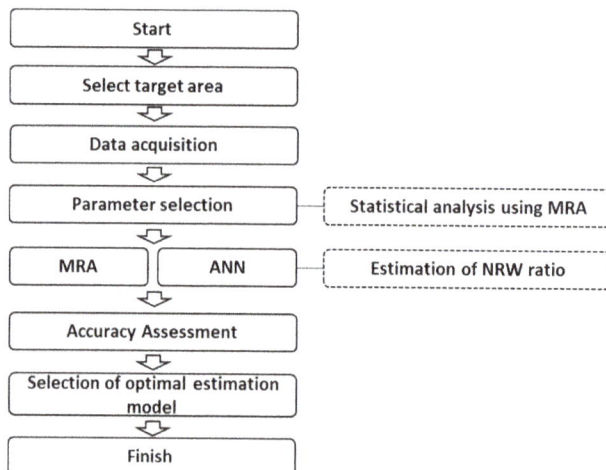

Figure 2. Study Diagram. Multiple regression analysis (MRA); artificial neural network (ANN).

The procedure for NRW ratio estimation utilizes MRA and ANN. The ratio results calculated from the selected parameters and measured value must be compared for analysis of the relations between used parameters and NRW. For this, an accuracy assessment method can be used. The mean absolute error (MAE), mean squared error (MSE), percent of bias (PBIAS), goodness of fit (*G*) value, and determination coefficient (R^2) can be commonly used to compare measured and simulated values. Finally, the parameters that represent the simulated value most similar to the measured value is selected for application of ANN.

Figure 1 shows the ANN structure for estimating the NRW ratio in a water distribution system. The input layer consists of physical and operational parameters and a bias layer. The bias layer can be considered a parameter that corrects the calculation results of the NRW ratio by ANN so that the measurement results and deviation will be reduced. The number of hidden layers is determined through trial and error, and the optimal number of neurons gives a highly accurate result. The optimal number of neurons as being composed of within 2n which was suggested by Heaton, J., where n is the number of independent variables [26].

4. Application to Study Area

4.1. Description of Data Collection and Parameters

The target area for this study was Incheon, Korea's third-biggest city. The data covered the status of the area, waterworks facilities and operational status, as well as water supply indicators of the Incheon Waterworks Basic Plan [27]. In addition, data on analysis of water pipe networks and simulation data of water distribution systems were collected [28].

The city's 367 district metered areas (DMAs) were selected for technical diagnosis analysis and a general technical diagnosis was conducted on 330 completed DMAs, as shown in Table 2 and Figure 3. In Figure 3, DMAs are colored gray and unblocked DMAs are colored white.

Table 2. Status of Incheon's DMAs.

Classification	Total
DMA (District metered area)	278 (76%)
From DMA to unblocked DMA	52 (14%)
Unblocked DMA	37 (10%)
* Total	367 (100%)

Note: * Source: Waterworks Headquarters Incheon Metropolitan City [28].

In order to apply the methodology to study area for estimating the NRW ratio, 173 DMAs were selected for estimation analysis and 194 DMAs excluded from the data for being unfinished, non-operational or running abnormally among the 367 DMAs.

Six selected parameters were used as independent variables for estimating the NRW ratio. Among the parameters reflecting physical characteristics selected were mean pipe diameter, pipe length per demand junction, amount of water supply per demand junction and deteriorated pipe ratio; the demand energy ratio and the number of leaks were selected as operational parameters.

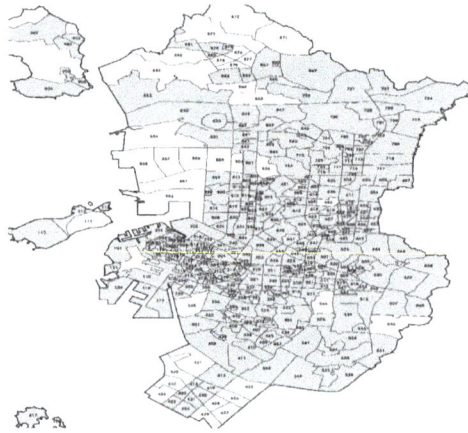

Figure 3. Water Distribution System of DMAs in Incheon [28].

The demand energy ratio is calculated by hydraulic pressure and demand in water distribution systems from the research of Jo (2017) [24]. The data used for estimating the NRW ratio are shown in Table 3.

Table 3. Selected parameters for estimating NRW ratio.

Classification	Parameter	Data Collection
Operational parameter	Demand energy ratio No. of leaks	Simulated data Measured data
Physical parameter	Mean pipe diameter Pipe length/demand junction Water supply quantity/demand junction Deteriorated pipe ratio	Designed data Designed data Measured data Designed data

Selected parameters were determined by previous research on parameter classification systems in water distribution networks [25]. These parameters were related with NRW and verified by statistical analysis between the parameters and NRW. Jang (2017) suggested various parameters, simulated estimating the NRW ratio, and determined the six parameters with high correlation and accuracy for estimating the NRW ratio [25]. Data on the selected parameters were collected in the target area. The NRW ratio, a dependent variable, is based on measured data in consideration of revenue water and effective and ineffective NRW in 2014, as shown in Figure 4.

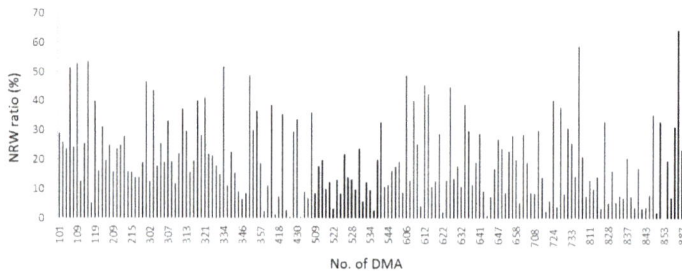

Figure 4. Data Status of NRW ratio in Incheon's DMAs.

The average NRW ratio of the 173 DMAs was 19.9%, the minimum NRW ratio was 0.2% (DMA No. 853), and the maximum NRW ratio was 64.3% (DMA No. 886). Figures 5–10 are the data of each independent variable.

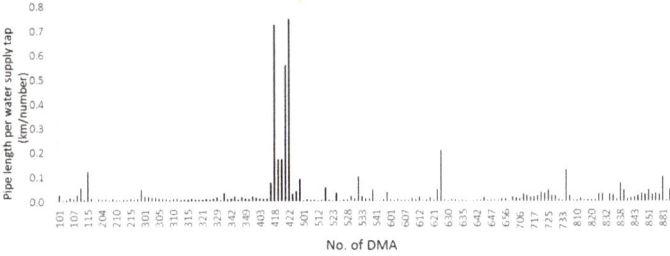

Figure 5. Data of Pipe Length per Demand Junction in Incheon's DMAs.

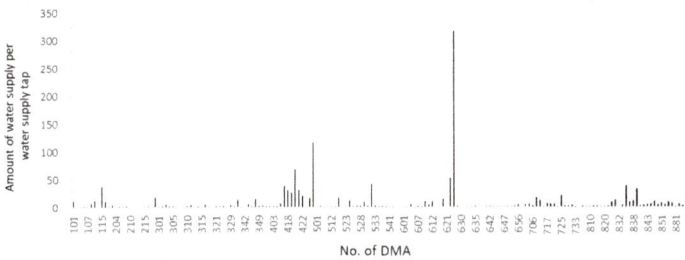

Figure 6. Data on Volume of Water Supply per Demand Junction in Incheon's DMA.

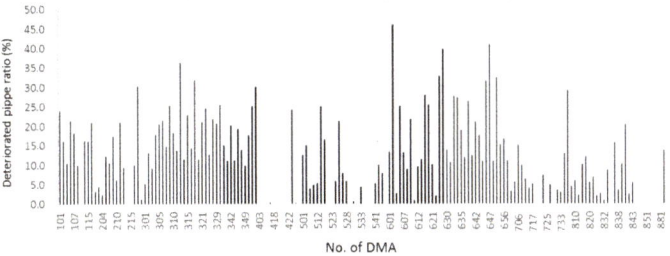

Figure 7. Data on Deteriorated Pipe Ratio in Incheon's DMAs.

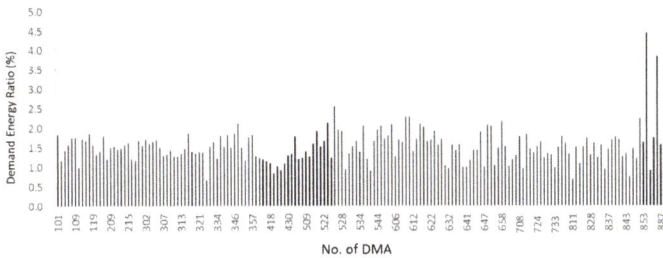

Figure 8. Data on Demand Energy Ratio in Incheon's DMAs.

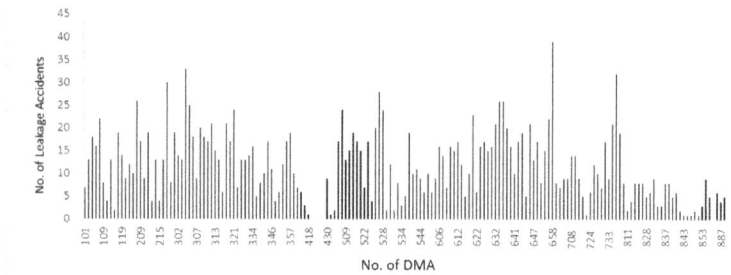

Figure 9. Data on No. of Leaks in Incheon's DMAs.

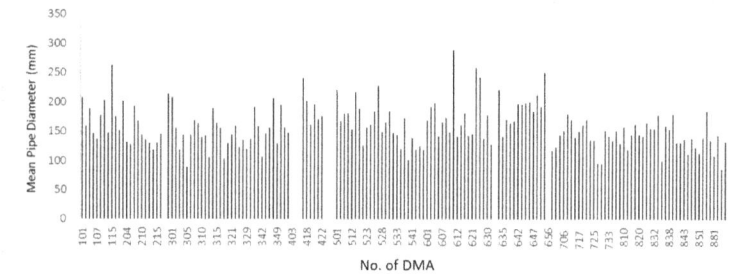

Figure 10. Data on Mean Pipe Dimeter in Incheon's DMAs.

The average value of the pipe length per demand junction (km/no. of junctions) is 0.0336; the maximum value is 0.75 (DMA No. 418) at the beginning of the No. 400 DMA located in industrial district, and; the number of demand junctions is lower than in residential district. Compared to other DMAs, pipe length of industrial district is longer than other districts for water delivery in each DMA.

The highest volume of water supply per demand junction was 318 in DMA No. 627 in an old industrial district. It was significantly higher than the average (9.1). The other district were mainly distributed with low value as less than 50 and there were some high values depending on DMA characteristics.

On average, pipelines more than 20 years old had an average NRW ratio of 11.3%; DMA No. 602 in the old industrial district had a ratio of 46.1%, significantly higher than other districts.

4.2. Result of MRA for Estimating NRW Ratio

The coefficient of equation for MRA was minus 0.048 for water supply quantity per demand junction, 0.284 for number of leaks, 0.347 for the deteriorated pipe ratio, 1.709 for the demand energy ratio, minus 0.064 for mean pipe diameter and 10.456 for pipe length per demand junction. The constant was considered a little high at 20.488, and three parameters were statistically satisfactory because the significance probability was lower than 0.05, but the other parameters were unsatisfactory. Results of MRA are shown in Table 4 and the calculated multiple regression equation is shown as Equation (4).

$$\begin{aligned} \text{NRW ratio (\%)} = {} & 20.488 - 0.048 \text{ water supply quantity per junction} + 0.284 \text{ number of leaks} \\ & + 0.347 \text{ deteriorated pipe ratio} + 1.709 \text{ demand energy ratio} - 0.064 \text{ mean pipe diameter} \\ & + 10.546 \text{ pipe length per junction} \end{aligned} \quad (4)$$

Figure 11 is the scatter plot graph. All original parameters were selected and applied to MRA. In the graph, the R^2 of 0.15 shows a low correlation with the real measured NRW ratio.

Table 4. Results of Multiple Regression Analysis.

Classification	Unstandardized Coefficients	Significant Probability
Constant	20.488	0.000
Water supply quantity per demand junction	−0.048	0.251
No. of leaks	0.284	0.074
Deteriorated pipe ratio	0.347	0.003
Demand energy ratio	1.709	0.454
Mean pipe diameter	−0.064	0.036
Pipe length per demand junction	10.456	0.379

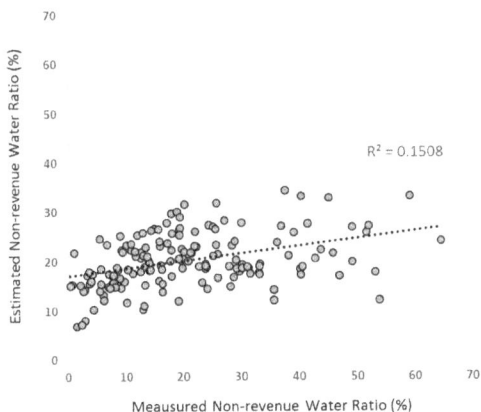

Figure 11. Scatter Plot of NRW Ratio Using MRA Results.

In order to calculate the multiple nonlinear regression for Equation (3), which was calculated through multiple regression analysis, a new equation was derived by introducing the exponential concept into each independent variable.

In order to determine the optimal formula, calculations of 1400 iterations were performed. The final parameters of nonlinear equation were calculated by repeating the calculation until the sum of squares error showed the smallest value. Finally the multiple nonlinear regression formula was derived, as shown in Equation (5).

$$
\begin{aligned}
\text{NRW ratio (\%)} = {} & 18.316 + 68.6 \text{ water supply quantity per junction}^{-0.0240} \\
& + 0.038 \text{ number of leaks}^{1.511} + 5.945 \text{ deteriorated pipe ratio}^{0.274} + 0.038 \text{ demand energy ratio}^{4.368} \\
& - 6.065 \text{ mean pipe diameter}^{0.345} - 32.523 \text{ pipe length per junction}^{-0.059}
\end{aligned} \tag{5}
$$

The result of the comparison between actual and simulated values is shown in Figure 12. As shown in Figure 12, R^2 is 0.19, which is approximately 21% higher than that of the multiple regression of Equation (3).

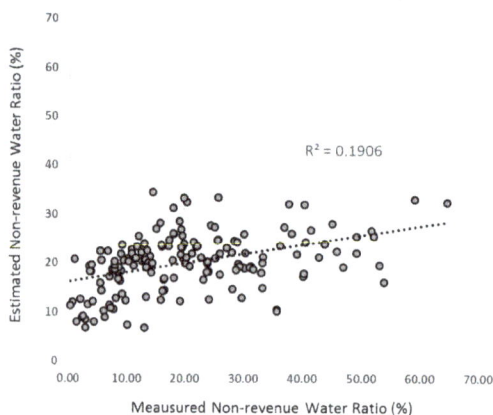

Figure 12. Scatter Plot of NRW Ratio Using Nonlinear MRA Results.

4.3. Result of ANN for Estimating NRW Ratio

The simulation condition for calculating the NRW ratio through ANN used the deteriorated pipe ratio, mean pipe diameter, water supply quantity per demand junction, demand energy ratio, pipe length per demand junction and number of leaks including both physical and operational parameters.

The number of hidden layers was capped at 15. To determine the optimal number of neurons in hidden layers, the method of trial and error was used for estimating the NRW ratio. When 12 neurons in hidden layers were selected for the optimal ANN model, they showed highest R^2 value than other the number of neurons applied case.

Figure 13 is a graph showing the comparison between the observation and simulation values of the NRW ratio under the 12 neuron condition in the hidden layer, which are derived from the optimal results among the 15 neurons.

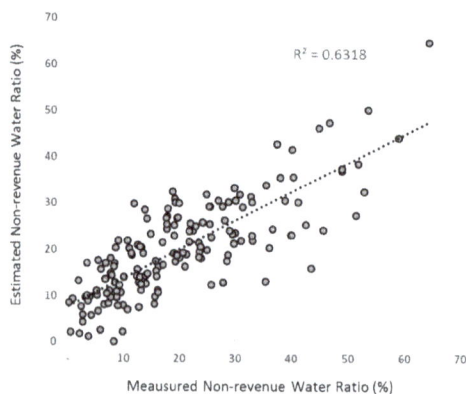

Figure 13. Scatter Plot of NRW Ratio Using ANN Results.

4.4. Accuracy Assessment for Analyzing Application Method

To evaluate the accuracy of the multiple regression equations proposed in the previous study and the results of ANN found by this research, an error ratio analysis was performed to evaluate the difference between the actual and model values. Accuracy analysis can be estimated by comparing the actual value with the value generated by the model.

This meant the use of the mean absolute error (MAE), the mean square error (MSE), and the percent of bias (PBIAS), which evaluates the bias of the estimation result, and the G-value prediction method. The calculation method of each equation is shown in Equations (6)–(9), and the comparison between the actual and model values can be more accurately evaluated through regression analysis.

$$\text{MAE} = \frac{1}{n} \sum_{i=1}^{n} [|z(x_i) - \hat{z}(x_i)|] \tag{6}$$

$$\text{MSE} = \frac{1}{n} \sum_{i=1}^{n} [z(x_i) - \hat{z}(x_i)]^2 \tag{7}$$

$$\text{PBIAS} = \frac{1}{n} \sum_{i=1}^{n} (z(x_i) - \hat{z}(x_i)) \tag{8}$$

$$G = \left(1 - \frac{\sum_{i=1}^{n} [z(x_i) - \hat{z}(x_i)]^2}{\sum_{i=1}^{n} [z(x_i) - \bar{z}]^2}\right) \times 100 \tag{9}$$

If MAE and MSE are smaller, the estimated value is more accurate. If PBIAS is close to zero, the estimation result represents less bias. If the G value is 100, it is a perfect estimation. If the G value is negative, this indicates lower reliable than using the average of data values as 10a predictor.

Table 5 shows the results of the accuracy assessment for the estimated results of MRA and ANN that selected optimal the number of neurons as 12.

Table 5. Results of Accuracy Assessment. Mean absolute error (MAE); mean squared error (MSE); percent of bias (PBIAS); goodness of fit (G).

Classification	Measured NRW Ratio	Multiple Regression Analysis	Non-Linear Multiple Regression Analysis	ANN (12 Neurons)
Sum.	3279	3279	3279	3297
Ave.	20.1	20.1	20.1	20.2
Standard D	-	23.7	23.5	21.9
MAE	-	10.0	9.9	6.2
MSE	-	146.8	139.9	63.7
PBIAS	-	0.0	0.0	−0.1
G	-	19.5	23.2	65.1

The ANN model with 12 neurons in the hidden layer was closest to the actual measured value. The condition using ANN satisfied all accuracy evaluation items except PBIAS and average value compared with MRA.

5. Conclusions

In this study, estimation of the NRW ratio using an ANN and MRA were conducted with specific parameters affecting the frequency of leaks in water distribution systems. Accuracy assessment was used to compare the selection of the optimal model between the ANN and MRA, and the following conclusions were drawn through the research.

Based on the results of the previous study, calculation of the NRW ratio is recommended by using MRA, which is obtained from the physical and operational parameters in water distribution networks. This study tried to use an ANN for estimating the NRW ratio, then compared the results of MRA and the ANN. An accuracy assessment showed that the ANN model had higher prediction accuracy than that of MRA.

A methodology has been developed for estimating the NRW ratio using an ANN with the main parameters of water distribution systems. When an ANN was used, the accuracy of NRW ratio estimation was higher than under the previous method of MRA. So when it is difficult to measure the NRW ratio and use MRA in a DMA, the ANN model is recommended for estimating the NRW ratio using the main parameters of water distribution systems.

When an ANN is used to predict the NRW ratio, finding the significance of the parameters is possible because it calculates the optimized model considering the interconnection between mutual parameters, despite the condition being statistically insignificant. The optimal number of neurons is also determined in the range of 2n when the number of independent parameter is n.

MRA is the most widely used method to predict the NRW ratio, and formulas vary according to regional characteristics and used parameters. If the volume of available data is sufficient, the ratio's prediction using an ANN is recommended for use in analysis of water distribution systems.

The use of the ANN model is expected to lead to the rehabilitation and improvement of water distribution systems and the optimal operation of water supply facilities for DMA construction.

Author Contributions: Dongwoo Jang performed the data analysis and modeling simulation. Gyewoon Choi contributed suggestion of methodologies and review the paper.

Conflicts of Interest: The authors declare that they have no conflicts of interest.

References

1. Rizzo, P. Water and Wastewater Pipe Nondestructive Evaluation and Health Monitoring: A Review. *Adv. Civ. Eng.* **2010**, *2010*, 818597. [CrossRef]
2. Saldarriaga, J.G.; Ochoa, S.; Moreno, M.E.; Romero, N.; Cortes, O.J. Prioritized Rehabilitation of Water Distribution Networks Using Dissipated Power Concept to Reduce Non-Revenue Water. *Urban Water J.* **2010**, *7*, 121–140. [CrossRef]
3. Engelhardt, M.O.; Skipworth, P.J.; Savic, D.A.; Saul, A.J.; Walters, G.A. Rehabilitation Strategies for Water Distribution Networks: A Literature Review with a UK Perspective. *Urban Water* **2000**, *2*, 153–170. [CrossRef]
4. Halhal, D.; Walters, G.A.; Ouzar, D.; Savic, D.A. Water Network Rehabilitation with a Structured Messy Genetic Algorithm. *J. Water Resour. Plan. Manag.* **1997**, *123*, 137–146. [CrossRef]
5. Frauendorfer, R.; Liemberger, R. *The Issues and Challenges of Reducing Non-Revenue Water*; Asian Development Bank: Mandaluyong, Philippines, 2010.
6. Alegre, H.; Hirner, W.; Baptista, J.M.; Parena, R. *Performance Indicators for Water Supply Services*; IWA Publishing: London, UK, 2000.
7. Lambert, A.O.; Hirner, W.H. *Losses from Water Supply Systems: Standard Terminology and Recommended Performance Measures*; IWA the Blue Pages, International Water Association: London, UK, 2000; pp. 1–13.
8. Lambert, A.O. *International Report on Water Losses Management and Techniques*; Water Science and Technology Water Supply; IWA Publishing: London, UK, 2002; Volume 2, pp. 1–20.
9. Liemberger, R. Do You Know How Misleading the Use of Wrong Performance Indicators can be? In Proceedings of the IWA Managing Leakage Conference, Nicosia, Cyprus, 20–22 November 2002.
10. Park, S.W.; Kim, T.Y.; Lim, K.Y.; Jun, H.D. Fuzzy Techniques to Establish Improvement Priorities of Water Pipes. *J. Korea Water Resour. Assoc.* **2011**, *44*, 903–913. [CrossRef]
11. Park, Y.S. A Study on Long Term Replacement and Maintenance Plan for Multi-Region Water Pipelines Considering Economics. Master's Thesis, Seoul National University, Seoul, Korea, 2014.
12. Park, I.C.; Kwon, K.W.; Cho, W.C.; Cho, K.H. Study on the Decision Priority of Rehabilitation for Water Distribution Network Based on Prediction of Pipe Deterioration. In Proceedings of the Korea Water Resources Association Conference, Jeju, Korea, 18–19 May 2006; pp. 1391–1394.
13. Mukundi, M.J. Determinants of High Non-Revenue Water: A Case of Water Utilities in Murang' A County, Kenya. Master's Thesis, Kenyatta University, Kenyatta, Kenya, 2014.
14. Shilehwa, C.M. Factors Influencing Water Supply's Non Revenue Water: A Case of Webuye Water Supply Scheme. Master's Thesis, University of Nairobi, Nairobi, Kenya, 2013.

15. Winarni, W. Infrastructure Leakage Index (ILI) as Water Losses Indicator. *Civ. Eng. Dimens.* **2009**, *11*, 126–134.

16. Wyatt, A.S. *Non-Revenue Water: Financial Model for Optimal Management in Developing Countries*; RTI Press: Amman, Jordan, 2010.

17. Jung, J.J. The Primary Factor of Management Evaluation Indicators for Local Public Water Supplies & Suggestion of Alternative Evaluation Indicators. *J. Korean Policy Stud.* **2012**, *12*, 139–159.

18. Lambert, A.O.; Brown, T.G.; Takizawa, M.; Weimer, D. A Review of Performance Indicators for Real Losses from Water Supply Systems. *J. Water SRT Aqua* **1999**, *48*, 227–237.

19. Shinde, V.R.; Hirayama, N.; Mugita, A.; Itoh, S. Revising the Existing Performance Indicator System for Small Water Supply Utilities in Japan. *Urban Water J.* **2013**, *10*, 377–393. [CrossRef]

20. Gwak, J.M. *Research and Statistical Analysis*; Informa: London, UK, 2013.

21. Haykin, S. *Neural Networks: A Comprehensive Foundation*; Macmillan: New York, NY, USA, 1994.

22. Kwon, S.H.; Lee, J.W.; Chung, G.H. Snow Damages Estimation Using Artificial Neural Network and Multiple Regression Analysis. *J. Korean Soc. Hazard Mitig.* **2017**, *17*, 315–325. [CrossRef]

23. Park, C.S. A Case Study on Establishment of Block System for the Increase of Revenue Water in Distribution Systems. Master's Thesis, Chonnam National University, Gwangju, Korea, 2014.

24. Jo, H.G. Study on Influence Factors of Non-revenue Water for Sustainable Management of Water Distribution Networks. Ph.D. Thesis, Incheon National University, Incheon, Korea, 2017.

25. Jang, D.W. Estimation of Non-Revenue Water Ratio Using PCA and ANN in Water Distribution Systems. Ph.D. Thesis, Incheon National University, Incheon, Korea, 2017.

26. Heaton, J.T. *Introduction to Neural Networks with Java*; Heaton Research, Inc.: London, UK, 2005.

27. Waterworks Headquarters, Incheon Metropolitan City. *Basic Plan of Waterworks Maintenance in Incheon*; Incheon Metropolitan City: Incheon, Korea, 2015.

28. Waterworks Headquarters, Incheon Metropolitan City. *Technical Diagnostics Report for Re-Establish Basic Plan of Waterworks Maintenance in Incheon Water Distribution Network*; Incheon Metropolitan City: Incheon, Korea, 2015.

water

MDPI

Review

Building a Methodology for Assessing Service Quality under Intermittent Domestic Water Supply

Assia Mokssit [1,2,*], **Bernard de Gouvello** [1,3], **Aurélie Chazerain** [2], **François Figuères** [2] and **Bruno Tassin** [1]

[1]   Laboratoire Eau Environnement et Systèmes Urbains, Université Paris-Est Créteil, Ecole des Ponts ParisTech, AgroParisTech, 6-8 Avenue Blaise Pascal, F 77420 Marne la Vallée CEDEX 2, France; bernard.de-gouvello@enpc.fr (B.d.G.); bruno.tassin@enpc.fr (B.T.)

[2]   SUEZ Group, Tour CB21, 16, Place de l'Iris, 92040 Paris La Défense CEDEX, France; aurelie.chazerain@suez.com (A.C.); francois.figueres@suez.com (F.F.)

[3]   Centre Scientifique et Technique du Bâtiment, 84 Avenue Jean Jaurès, F 77420 Marne la Vallée CEDEX 2, France

*   Correspondence: assia.mokssit@enpc.fr; Tel.: +33-781-800-819

Received: 20 July 2018; Accepted: 24 August 2018; Published: 30 August 2018

Abstract: This document proposes a methodology for assessing the quality of water distribution service in the context of intermittent supply, based on a comparison of joint results from literature reviews and feedback from drinking water operators who had managed these networks, with standards for defining the quality of drinking water service. The paper begins by reviewing and proposing an analysis of the definition and characterization of intermittent water supply (IWS), highlighting some important findings. The diversity of approaches used to address the issue and the difficulty of defining a precise and detailed history of water supply in the affected systems broadens the spectrum of intermittency characterization and the problems it raises. The underlined results are then used to structure an evaluation framework for the water service and to develop improvement paths defined in the intermittent networks. The resulting framework highlights the means available to water stakeholders to assess their operational and management performance in achieving the improvement objectives defined by the environmental and socio-economic contexts in which the network operates. Practical examples of intermittent system management are collected from water system operators and presented for illustration purposes (Jeddah, Algiers, Port-au-Prince, Amman, Cartagena, Barranquilla, Mexico, Cancun, Saltillo, Mumbai, Delhi, Coimbatore ...).

Keywords: intermittent water supply; water service quality; 24 × 7; water accessibility; water affordability; water availability

1. Introduction

Access to safe and affordable drinking water is considered by the United Nations as essential to the full exercise of a human being's right to life [1]. Yet, despite increases in global coverage, more than 663 million people in the world still lacked access to improved drinking water in 2015 [2], and approximately 2.1 billion people worldwide lack access to safely managed drinking water [2]. Also, with climate change, demographic growth, and urban sprawl that characterize our era, the water distribution situation is a challenging issue.

The above factors impact the availability of resources and the need for water. At the city level, this translates into a shortfall of water injected into the distribution network, faced with a growing demand for water from the population. However, supply increase is usually costly and not always possible, and on the other side, demand management is not entirely within the operator's action scope, as it necessitates customer involvement and investment in actions like sanitary devices replacement.

Distribution networks are usually the weak point of the whole system, and may be optimized to improve the situation [3].

While we can't entirely control the demand or the resource, advances in water supply play a key role in improving the situation. In this regard, simulating water distribution systems behavior with optimization purposes aims at effectively enabling human beings to access the sufficient water they have a right to, in order to meet their basic needs.

In reality, not all distribution systems fulfill this role in the same way. Due to various constraints, rendering water available according to the consumer's need is not always achieved [4], water in particular is not always delivered continuously and its quality is significantly impacted. The service is subject to planned or unplanned cuts. This supply regime, which we will initially call intermittent distribution, has been *de facto* put in practice by some water distributors in recent decades. There are 1312 billion people from a total of 2910 billion with access to piped water that are affected by intermittency worldwide [5], and an estimated one-third of piped water supplies in Africa and Latin America and more than half in Asia supply water intermittently [6]. The magnitude of the population affected has also led the International Water Association (IWA) to create a Specialist Group on Intermittent Water Supply (IWS) in 2017.

Intermittency causes technical malfunctions, but also inequity in water access, pricing intricacies and health problems for consumers. It generates or aggravates water leakage and financial losses for the operator in contexts where water is already scarce [7], and unnecessary waste and deterioration of water quality even when the resource is abundant.

Continuous distribution is all the more desirable as the distribution networks are originally designed for a permanent supply of pressurized water. Otherwise, some of these networks are oversized due to intermittency and the need to distribute considerable volumes of water in a few hours.

However, switching to continuous distribution is neither always possible nor easy to achieve in the short term, due to one or several reasons such as: resource unavailability, infrastructure deterioration, or specific political, economic, or cultural issues. Intermittency is then often established over long periods of time, and optimized management of this supply regime becomes necessary.

The optimization of supply quality in these conditions requires an initial detailed characterization of the situation accompanied by an identification of possible and desired improvement paths. This paper gives a review on intermittency and an analysis of the objectives for improvement of water service quality. This work is a basis for the establishment of the definition of a resilient operation, maintenance and management of intermittent water supply, when the transition to continuous supply is not possible in the short term.

2. Approach and Documentary Basis

This study presents the construction of a methodology for evaluating the quality of water distribution service in the context of intermittent networks. The first objective is to supplement the results of the scientific literature by feedback from practical field experience, in order to define a technical and operational characterization of intermittency. These results are then compared with commonly accepted norms and standards in water supply. An analysis of the possibilities for improvement, based on the study of the gap between these two states (real and desirable), is finally formulated.

The study is therefore carried out in three main stages (see Figure 1):

1. Complementary analysis of the feedback from the experience of operators having managed water networks in intermittent regime, and the results of a documentary review for the qualification of intermittency (First literature review).
2. Literature review on the quality of drinking water distribution services: International standards and minimum requirements to meet the human right to water and the SDG (Sustainable Development Goals) targets (Second literature review).
3. Extraction of quality of service qualification parameters that make sense in the context of intermittency, through a cross analysis between the characterization of intermittency and the

universal norms and standards defining water service quality, and global IWS Indicators framework construction.

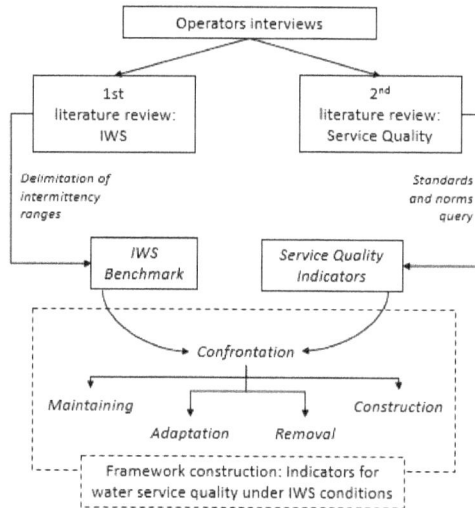

Figure 1. Approach Framework.

The first literature review (IWS literature review) consisted of identifying different aspects of intermittency, its causes, consequences, problems, and categorizations, but also its modelling and design. More than 70 academic journal articles published up to March 2018 were consulted on the subject on the World Wide Science Database, SciELO, Google Scholar Database, and Scopus. The key search expressions used were a combination of the following terms: [{intermittent (77)/irregular (8)/unreliable (22)}/{conversion to 24/7 (2)/continuous (5)}] − [domestic (5)/household (8) − water (97) − distribution (29)/supply (71)].

It is interesting to note that the literature collected on intermittency is relatively recent and scarce despite the size of the population affected by this phenomenon (cf. Figure 2). However, the temporal distribution of the articles tends towards a general increase which is a sign of a growing questioning, one that might gain even more scope with the creation of the IWA specialist group.

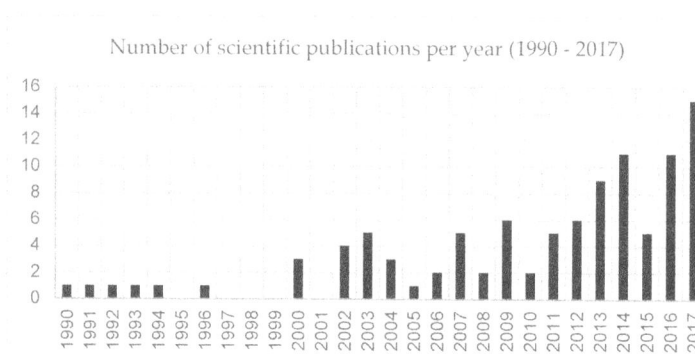

Figure 2. Temporal repartition of the references of the Intermittent Water Supply (IWS) literature review.

The objective of this review was to identify a definition of this regime and its extent, and to establish a global and aggregated categorization of the causes and consequences that could address the various aspects that characterize intermittency. This analysis has required the detection and distinguishing of features that are similar, in order to regroup them in more global categories.

The relevance of each characteristic feature was determined by a prioritization established by the operators who were interviewed. The reason behind this procedure is that what happens in the field is not necessarily reported or published, and therefore some aspects of intermittency may not have been dealt with in the literature. The purpose of the interviews was thus to report on the operators' findings following their experience with intermittent network management, in order to complete the information gathered from the literature, in addition to extracting their view on the issues put forward in literature for a more operational and field point of view.

The series of interviews were conducted with water service agents, experts and managers in several cities in Latin America, Asia, and Africa, those regions being the most impacted by IWS worldwide [5,6].

The people interviewed are all operators or experts who have worked on drinking water networks under intermittent conditions. A constant two-way flow between interviews and literature results was necessary to contextualize the networks and assess their generalization potential, and to complete and compare the information collected.

Twelve operators were interviewed, 10 who had a relevant experience in SUEZ's Business Units under IWS all around the world (Algiers (Algeria), Jeddah (Saudi Arabia), Amman (Jordan), Port-au-Prince (Haiti), Cartagena, Barranquilla (Colombia), Mexico, Cancun, Saltillo (Mexico), Mumbai, Delhi, Coimbatore (India), . . .), and two from Jeddah's and Jordan's National Water companies.

The list of the interviewees is presented in Table 1:

Table 1. List of interviewees.

Name	Occupation Relevant to IWS
Juan MATEOS IÑIGUEZ	Former Engineer in Barranquilla and Cartagena (1996–2000)
Philippe CARTON	Technical Director of the Port-Au-Prince and Haiti contract since 2014, former Director of the Amman management contract (2001–2003)
David DUCCINI	Former Technical Director at PALYJA, Jakarta (1999–2003), and SEAAL, Algiers (2006–2008)
Didier SINAPAH	Former DMA/DMZ manager in Jeddah (2008–2010)
François FIGUERES	Former technical director in Mumbai (2014–2017)
Fayçal TAGHLABI	Former engineer at Port-au-Prince (between October 2013 and March 2014)
Diane D'ARRAS	Intervention in Port-au-Prince in 2013, present at the beginning of the Buenos Aires contract (1993–1994)
Pascale GUIFFANT and Aymeric BAJOT	Support to SUEZ's operational and commercial entities in the implementation of societal programs for access to water and sanitation in informal settlements (Mumbai, Algiers, Jakarta, Casablanca, Haiti and other cities) (2014–2018)
Nicolás MONTERDE ROCA	Technical Manager in "Aguas de Saltillo" (2011–2017). Responsible for drinking and waste water networks O&M. Planning Manager at "Aguas de Saltillo" (2017–Present) in Mexico
Abdulrahman ALSEHRI	Former Water Supply Manager (2012–2015) and former Transmission manager (2009–2012) in Jeddah
Jiries DABABNEH	Technical Services Director of "Miyahuna", Amman (2008–2016)

Depending on operator availability, interviews were either conducted face-to-face or questionnaires were sent to individuals for completion (Jeddah, Saltillo, and Amman). The two types of interviews dealt with the following aspects:

- Experience with IWS systems: description of the network and the operator's role;
- Nature of intermittency: frequency, predictability . . . ;

- Causes: Reasons for intermittency deployment;
- History: Network's past regimes;
- Malfunctions: Problems related to intermittency and their ranking;
- Conversion to continuous supply: Interest in and attempts to switching to continuous supply;
- Modelling and available data: Use of existing and adapted models regarding the collected data;
- Stakeholders: Different actors and entities involved in IWS management.

The objective of this double analysis was to establish an overall characterization of intermittency. It entailed a technical definition of the regime and a characterizing pattern of its evolution, from its triggering elements to the various events it can generate. This result was obtained following an aggregated categorization of the main global causes of intermittency and their resulting consequences, by highlighting the multiple interrelations that characterize the evolution of service quality under IWS.

The resulting characterization of IWS was then confronted to the international standards for drinking water distribution, notably those in line with the Sustainable Development Goals that were agreed upon by all the United Nations Member States [8]. This stage required the second literature review where recent reference texts and reports of international organizations dealing with the issue of water supply were consulted (the United Nations, World Health Organization, UNICEF, IWA, etc.). The objective of this phase was to outline the existing criteria for evaluating a drinking water service, and their associated scales or ladders, and examine their relevance for IWS networks.

The standard criteria were either maintained, adapted, or suppressed in agreement with the characterization defined in the precedent stage, and some new IWS indicators were constructed accordingly. Scales were then established based on both the specificities of this flow regime and its potential for improvement, within intermittency and outside it.

This phase made it possible to produce a scaling method of drinking water networks under IWS, according to the level of service quality that characterizes them. This scaling method has been applied to the case of Jeddah and Saltillo, which are examples of intermittent networks whose service has improved without completely switching to continuous supply up to date. This final stage was carried out by examining the technical documentation provided by the water companies managing these two networks.

3. Intermittency Characterization

3.1. Definition of Intermittency

IWS encompasses a broad spectrum of practices with varying water delivery durations and patterns [9]. There is no comprehensive, shared, and unanimous definition of intermittent water supply in the literature, it is often linked to rationing, apportioning or restriction [10–14], or defined by the impact it has on the network [15,16], sometimes stressing its inadequacy [14]. However, a core first description can be extracted from proposals by different authors and summarized as follows: "Intermittent water supply is a piped water distribution service that provides water to users for less than 24 h in a day" [5,17–21].

This description is broadly echoed with slight variations and additional elements according to the authors. In particular, the notion of regularity is not unanimously agreed upon. For Gohil (2013), even under intermittent supply, water needs to be delivered at regular intervals throughout the day, such as a few hours in the morning or in the evening [22], whereas for Shrestha and Buchberger (2013), water supply is said to be intermittent when the water of a region is *regularly or irregularly* conveyed by pipeline networks for less than 24 h a day [23], where irregular intermittency refers to a supply arriving at unknown intervals within short time periods of no more than a few days [4].

The duration of supply that characterizes this type of water distribution is also a point of divergence in the literature. The temporal restriction is well established and emphasized everywhere [24,25], but some authors distance themselves from the less than 24 h in a day limitation, widening the spectrum to a supply to end users lasting anywhere from limited hours a day to only on a few days a week or less [4,7,26–28].

The frequency or periodicity with which water is supplied is not definite either, it is generally considered daily [29], although in extreme cases it lasts more than one day [30], and water has overall an "infrequent availability" [31]. In certain intermittent networks like Jeddah, Port-au-Prince, or Coimbatore, the adopted water distribution schedule is weekly or even monthly rather than day to day, the limitation in time is persistent but the supply times and their periodicity are diverse. In any case, the frequency of interruption in IWS must be persistent [31], intermittency differs from exceptional or accidental service interruptions, which otherwise cover more localized areas. Frequency of service interruptions is a usual service quality indicator that can be revised in the case of IWS networks.

If intermittency may seem like an adaptation to a circumstantial situation, it can also be deliberate. Supply interruption can be implemented to save water and/or energy [21,32,33], or for economic, political, or security factors [14].

All in all, the characteristics of intermittency apply to a piped distribution system, with a more or less regular supply that is limited in time. In addition, the distribution perimeter involved is delimited by the secondary and tertiary piped networks, from the source to their exit points (housing, kiosks, standpipes, etc.) [5]. Some consider the system as intermittent if all the network mains are not always full of water and pressurized [34]. From the operator's point of view, IWS is practically set up through cutoffs often established via valve operations at some sectors to allow adequate pressure in other parts of the network during that time [29,33,35,36]. From the receiver's end, the definition should include the global availability of water at the outlet points. A more technical definition considering both ends would then be to consider a system as intermittent when the residual pressure at the outlet points is not permanently above a given threshold. The threshold value must be appropriate to each network's characteristics. In the case of Jeddah's Network, for example, this value can be fixed to 0.5 bar, as per the Performance Indicators requested by the country's ministry of water and electricity for its management contract (2007), whereas in Coimbatore, the design parameter is for the pressure to be of at least 7 m of water column (0.69 bar) at the house connection [37].

The next two paragraphs result from the cross-analysis between the literature review and the operators' feedback, in order to display the most operationally significant dimensions of intermittency, with practical examples.

3.2. Causes of Intermittency

Despite the diversity of definitions, the majority of the authors and all the interviewed operators note the negative impact of intermittency on the quality of drinking water distribution, this aspect will be detailed in the next paragraph. However, a first question is why does this supply regime take place, or how else does it establish itself? Neither the operators interviewed nor the literature did pinpoint a clear-cut trigger that could have caused the supply to shift to an intermittent regime; rather, they evoked a general multifactorial constraining context at the resource or management level. These conditions are symptomatic of the general constraints to which drinking water networks are subjected today.

Water and energy scarcity, population density increase, and the dynamics of urban expansion in one side, combined with failures in governance and the lack of adequate planning and structural adaptation, in developing cities, make the task of providing basic urban services more complex [38,39]. It is generally a combination of several of these factors, sometimes with a contribution of the population's attitude towards water, that set intermittency off [7,40].

Totsuka, Trifunovic, and Vairavamoorthy (2004) put forward a structuration of these causes into three types of anomalies related either to resources, investment and infrastructure, or management [41]. In the following, we list examples of networks described by surveyed operators that are subject to one or the other of these anomalies.

The cities of Barranquilla and Cartagena in Colombia suffered from a lack of infrastructure to cope with the migratory flow caused by the states of insecurity in the South and Center of the country, which pushed a large part of the population to migrate to these two cities. This situation has led to a significant densification of the cities which, combined with a lack of planning, has caused an imbalance in water distribution. The problem was not a problem of resources, which exist in sufficient quantity, it was only a management problem aggravated by the pressure exerted by minorities, such as the rich or neighborhood leaders, on political will.

The master plan of the city of Cancun in Mexico was done in one shot without updating. As a result, the forecasts of this master plan have extremely underestimated the population growth of Cancun, closely linked to the region's thriving touristic activity, which has generated a demand for water five times greater than was predicted. In order to manage this situation, the peripheral districts of the city have been tactically put into intermittent mode to limit water theft due to illegal connections implemented by the residents of the suburban slums.

Finally, Mumbai is a striking example of a situation where resources and hydraulic capacity are sufficient but the network deterioration, customer behavior, uncontrolled urban densification (slums), and management liabilities induce service malfunction, since the city does not have a water resource problem as the ratio of water availability per person is 300 L per day while for the city of Barcelona, where water is delivered continuously, the dotation per person per day does not exceed 180 L.

When a water company is confronted with one of these anomalies, the adopted solutions can range from a continuous supply with low flows, with the consequent reduction of pressures on the network, to the use of recycled water or unconventional sources, and finally to an intermittent supply [42].

3.3. Problems and Risks Associated with Intermittency

All interviewed operators along with the literature stress the negative impact of intermittency on the network, and its impact on the efficiency of operations, demand management, and water supply in general [5,41]. The most frequently mentioned impacts further include:

- Health risks: Water quality and health hazard [6,9,28,43];
- Technical problems: Network wear [17,26,39]; Difficulty in detecting and repairing leaks [5];
- Economic issues: The cost of network wear; Metering and billing issues [44,45]
- Social and political problems: Illegal connections and other customers' coping strategies [10,39,46,47]; Inequity of water supply [7,10,48,49]; Water wastage [5,7];

And overall restrictions in the quality of supply [7] such as water unavailability, etc.

In what follows we will report on the various aspects of this adverse effect on the network, and other problems that have been identified by operators and in the literature. The last point—restriction in the quality of supply—will be explored in detail in the next part.

3.3.1. Health Risks

The interruption of supply is the main characteristic of intermittency; it causes low pressures and vacuum in the pipes for relatively long periods [5]. This makes them favorable to contamination as soon as there is a route of penetration of external substances and organisms, and as we will further detail, these networks are generally very leaky, there is a rather high risk of the existence of this route. Moreover, variations in pressure and velocity induce bacterial biofilm detachment and microbial regrowth within the pipes [16]. In addition, because it fails to deliver an adequate service to the consumers, namely available water when needed, intermittency leads the consumers to the use of complementary systems such as domestic water storage accommodations, like a water tank on the house roof, cistern, or other permanent deposit at home [10,18,44,46,47,50–52]. Because of its stagnation in domestic tanks or reservoirs, water quality is deteriorated again at the consumer's side [28,31,53–55], even if there is a regulation of the construction and maintenance of these thanks, this type of installation is private property. In practice, they are rarely built, controlled, and maintained according to standards that take into consideration prevention against contamination.

This aspect was noted by several operators in Amman, Mumbai, New Delhi, and Port-au-Prince.

High risks of contamination create or intensify major health risks [56], and some intermittent cities are experiencing key health problems [9,11,53,57,58], as is the case for the population of Port-au-Prince, which suffers from the spread of cholera, necessitating domestic water treatment, in order to be reduced [59].

Finally, IWS also presents risks in terms of hygiene behavior deterioration. Rationing reduces water consumption, and thus washing frequency as well, along with the increase of water sharing among family members, inducing additional health risks [14]. Figure 3 is an illustration of the aforementioned health risks.

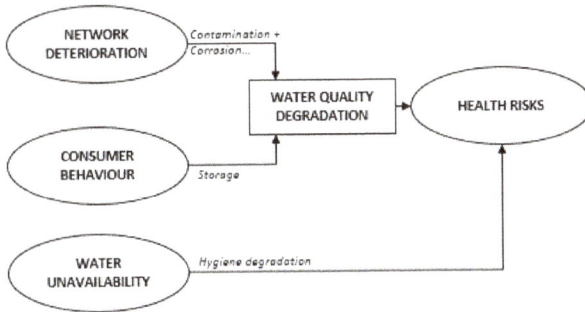

Figure 3. Factors inducing or aggravating Health Risks in IWS.

3.3.2. Technical

i   Network Deterioration

Intermittent systems experience more repetitive fluctuations in pipe pressures than continuous systems [48,60]. Furthermore, entrapped air bubbles may get overpressured following their infiltration to the pipes during the supply halt periods [24,61]. These two phenomena cause an acceleration in the wear of the pipes and joints, and thus an increase in their breakage rate [62]. This deterioration also affects equipment installed on the network, which are handled more frequently. All the operators interviewed testified to the degradation of the network that they had to manage, resulting in an increase in the breakage rate and maintenance costs in this type of network, Figure 4 illustrates the evolution of leakage in Hussein Dey, a 56 linear km Network pilot zone in Algiers before and after its transition from intermittent to continuous supply, the average number of reported leaks significantly decreases both on distribution mains and connections (cf. Table 2). This phenomenon was also confirmed by a study carried out in Cyprus where the modelling of the behavior of a network during and after intermittency, displayed its vulnerability following the implementation of IWS [17,22].

Table 2. Effect on intermittent supply on reported leaks (Algiers)—Adapted values from David Duccini's presentation at the Journées Techniques Innovation—SUEZ (2009).

Description	Number of Reported Breaks		
	Before	**After**	**% Decrease**
Mains	10.08 in 1 km	3.5 in 1 km	−65%
Service connections	3.27 in 1000	0.5 in 1000	−85%

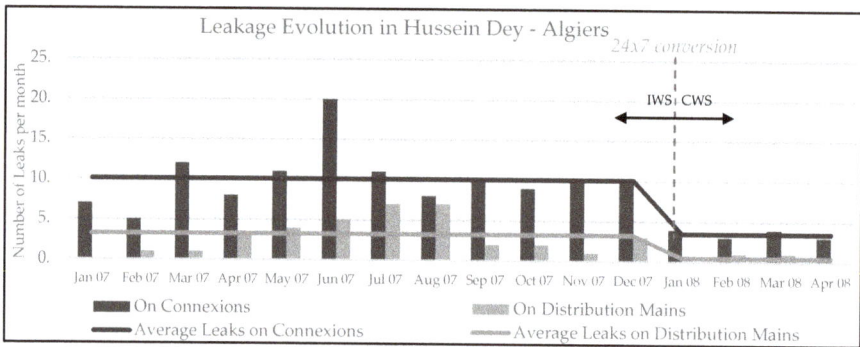

Figure 4. Leakage evolution in Hussein Dey—Algiers before and after the transition to continuous water supply (CWS—January 2008): Adapted chart from David Duccini's presentation at the Journées Techniques Innovation—SUEZ (2009).

ii   Increased Difficulty in Leak Detection and Repair

The more the network is degraded, the more there is room for leakage, and the more water is lost [62]. However, it should also be noted that intermittency makes it more difficult to detect leaks in general, whether they are its direct consequence or not [5], because the usual leak detection techniques (fixed and mobile pre-localization, acoustic detection, etc.) require a pressurized network, with the exception of Helium detection (tracer gas). And even in the latter, the repair cannot be validated on the spot due to the absence of pressure. This problem was noted in Amman in particular, where leaks that were supposed to be repaired reappeared once the network was put back under pressure. Figure 5 illustrates some factors generating or aggravating network deterioration and leakage under intermittency.

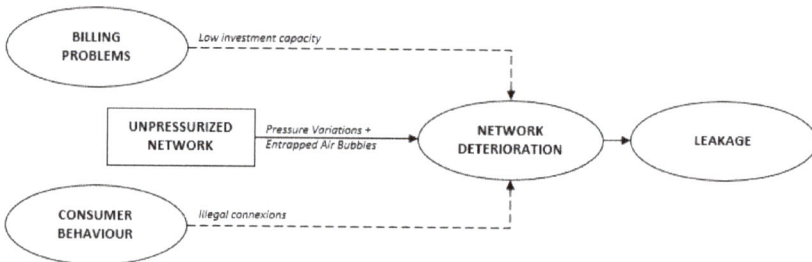

Figure 5. Factors inducing or aggravating Network degradation and leakage in IWS.

3.3.3. Economic Issues: Billing Problems Because of Meter Malfunctioning

Intermittency induces an interruption of service, usually associated with cycles of water filling and emptying of the pipes, and the expulsion of the air that was entrapped in the network. These conditions cause metering malfunctions readings [44,45] due to the following reasons:

- Firstly, the alternation of dry and wet conditions weakens the devices;
- Then the air movements in the pipes accelerate the deterioration of the meters' recording mechanism when air is driven out of the pipe by incoming water;
- And when the supply is switched off, it creates vacuum conditions reversing the meter registration [5].

These phenomena create difficulties for water suppliers in controlling water consumption and establishing fair pricing [63] (cf. Figure 6). All the more so, as these factors make consumers suspicious of the accuracy of their bills and less willing to pay, and can sometimes lead to criminal behavior, in Cancun for example, operators have observed a recurrent practice of meter sabotage from populations supplied in intermittent mode. However, it should be noted that this practice is more or less pronounced depending on the cultural sphere concerned and does not necessarily depend on intermittency, even if it may be aggravated or justified by the feelings of dissatisfaction and injustice that IWS generates.

Figure 6. Factors inducing Billing Problems in IWS.

These problems lead water utilities to question the relevance of installing meters for this type of supply, but their absence generates other complications such as the obligation to introduce flat-rate or roughly evaluated billing [63], and the difficulty of estimating network performance.

3.3.4. Social and Political Problems

i   Inequity of Distribution within the Network

In an intermittently supplied system, the hydraulic conditions associated with consumer behavior and strategic decisions result in inequitable water allocation [19,29,30,64–67].

On the one hand, the hydraulic structure of the network makes some points more advantageous hydraulically than others because of their elevation and their proximity to the production points [42,48,49], especially when the network is originally designed for continuous supply [61,68]; the flow is characterized by a water reception time that is not equal everywhere; this characteristic constitutes a problem when this reception time is longer than the supply duration. De Marchis et al.'s model of the filling process of the network of the city of Palermo (2010), estimated the water reception time for the most disadvantaged points of the network as superior to an hour, the resulting pressure was then too low to fill the users' tanks, a few of them were completely filled within 5 h, while others had to wait up to 8 h for the tank filling process to start [48]. Second, the hydraulic regime in intermittent networks is governed by a pressure-dependent demand, the volume of water flowing from the taps is dependent on the pressure in the network. As consumers leave their taps open, filling the tanks causes high peak flows, which generate impressive pressure losses in the network, this process amplifies the pressure difference between the different collection points in the network and increases inequity [69]. This situation may be aggravated by the use of suction pumps by some consumers [70]. This practice unbalances the supply even in the case of a continuous supply network, as is the case in Buenos Aires. When the availability of water in the network is limited, particularly under intermittent conditions, the described impact is even worse.

Finally, some neighborhoods or consumers may be favored by the municipal authorities or the operator because of their social status, political weight or ability to pay.

Figure 7 displays some of the factors contributing in the inequity of IWS.

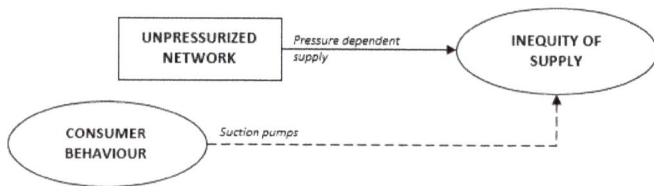

Figure 7. Factors inducing or aggravating Inequity of supply in IWS.

ii Water Wastage

Consumers in intermittent conditions may waste more water than those who can receive it permanently. In fact, they tend to keep their taps open to store as much water as possible, from fear of shortages, which incidentally causes occasional overflows in tanks. In addition, most consumers do not use all the stored water and this water will be tossed away and replaced by fresh water from the next supply window [27,41,63], especially when water tariffs are flat rate or state-subsidized. This situation is all the more paradoxical as intermittency is often initially caused by a lack of resources. This phenomenon is a salient example of the interweaving between the causes and consequences of intermittency, and the difficulty of escaping it [43] (cf. Figure 8).

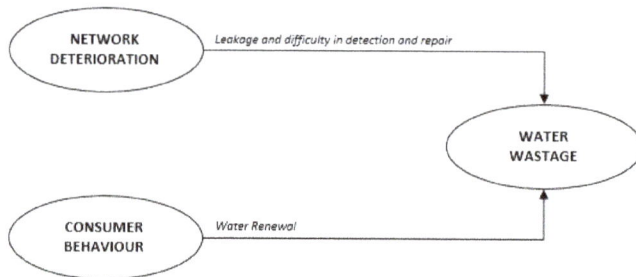

Figure 8. Factors inducing or aggravating water wastage in IWS.

iii Coping Strategies of Consumers

Over the years, consumers have begun to adopt various measures to overcome intermittency, but at a cost generally referred to as the "coping cost". A considerable coping cost is incurred to store backup water to ensure all-day availability [71].

The intermittent regime imposes on consumers the costs associated with additional facilities, such as storage tanks, pumps, alternative resource supply systems, and domestic treatment facilities (cf. Figure 9), generally afforded only by the wealthiest households. Poorer people who cannot pay for such accommodations either spend their time fetching water from alternative providers such as public taps or vendors, at a relatively high total price, or are compelled to reduce their consumption [10]. Coping costs are related in general to collecting, storing, pumping, treating, and purchasing water [46]. In Port-au-Prince in particular, when faced with the unavailability of water, consumers buy water from private tankers, which are more expensive than piped water. This argument is used by water utilities to encourage people to switch to continuous service, as is the case in New Delhi, where the operator highlights the potential savings achieved on these adaptation practices.

Moreover, slots where water is available are not always convenient for users. When water is needed but unavailable, people have to go to public taps, sometimes quite far away, and wait in long lines to collect it. In some countries, consumers' working hours are affected, resulting in lower overall productivity.

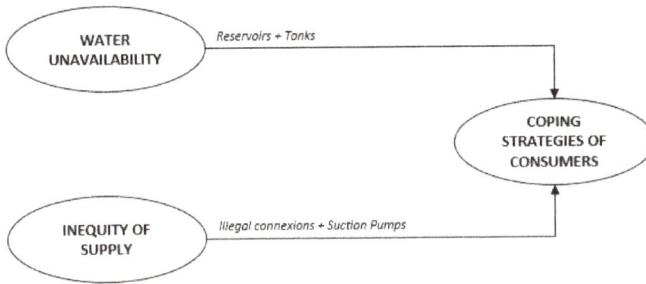

Figure 9. Some factors inducing or aggravating customer coping strategies in IWS.

Finally, faced with the unavailability of water, certain consumers illegally connect to the pipes that are pressurized in the network, and some even bribe the field agents who handle the valves in order to alter the distribution schedule in their favor. These practices disrupt the functioning of already fragile network and supply, some of them are represented in Figure 9.

3.4. Interrelation of IWS Causes and Consequences

Overall, intermittency is caused by an imbalance in the water supply/demand equation. On the one hand, there is the increase in demand caused by urban expansion and population growth, and the multiplicity of water uses, and on the other hand, supply is constrained by conditions of shortages that may be related to resources, infrastructure, mismanagement, or poor or static planning. The main characterization of intermittency is that its establishment aggravates the imbalance of the equation by engaging in a vicious circle of water service degradation. It degrades the network, which generates more leaks, more water losses, and more interventions and investment needs that add to the costs of the system operation and management.

A striking point that emerged from the interviews is the link between causes and consequences in the case of intermittency, problems caused by intermittence in some cases are causes of transition to this regime in others [5,40], Charalambous & Laspidou (2017) represent this interdependence as a regressive spiral of the water service situation, shown in Figure 10.

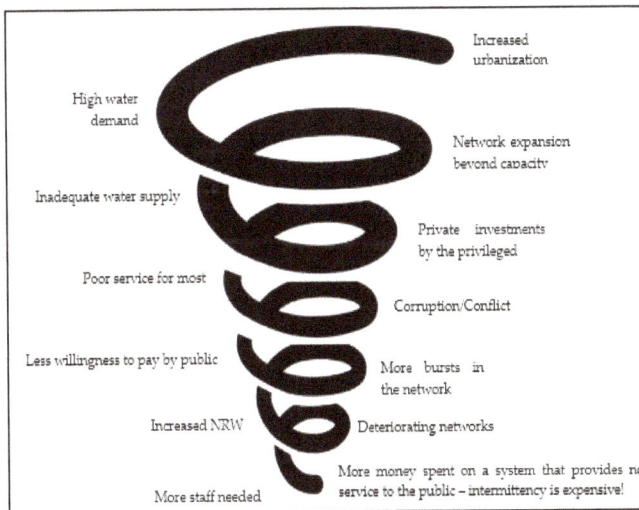

Figure 10. The downward spiral of IWS [5].

This spiral is a conceptual representation of the degradation of service under IWS, as it is broadly applicable for most intermittent networks, with few exceptions in the case of Jeddah for instance (5500 km of water network divided in 125 areas) where the origin of the transition to Intermittent Water Supply (IWS) is linked to the instability of water supply, which has pushed the populations to install individual storage tanks with capacities of 200 to 100,000 L, to compensate for the eventuality of a supply cut. The stages of the spiral can be more or less prominent and significant depending on the population and utility's financial resources. In Jeddah, for example, the cost of service is affordable by the majority of the consumers, and water services are highly state-subsidized (DS), thus there wasn't a particular stress on the consumers' willingness to pay.

We propose in the following (see Figure 11) an illustration of the sequence of relationships between the above causes of intermittency as defined by Totsuka et al. [41], combined with the related problems that we described in the precedent section. The multitude of interrelations (arrows) illustrates the service degradation expressed by the spiral and the vicious circle of intermittency, where the problems it generates fuel its origins, while stressing the non-linearity of the issue.

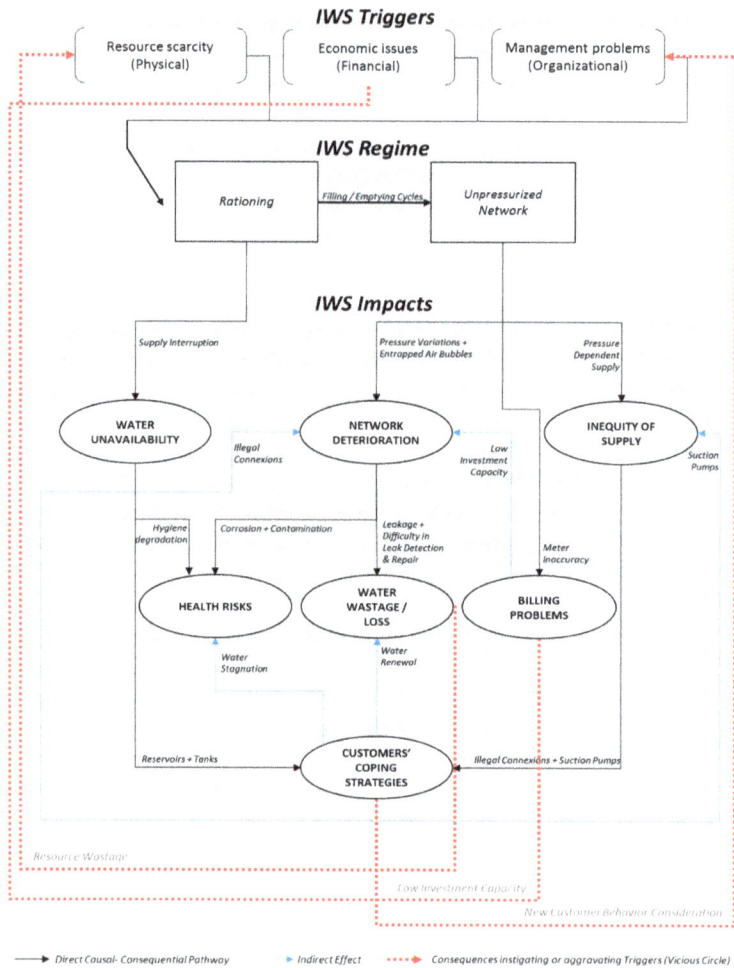

Figure 11. Global comprehensive process of the interrelation between the different dimensions of intermittency.

This illustration links all the negative results of intermittency (dotted arrows), directly or indirectly, to one of the imbalances that triggered this supply mode. Billing problems, for instance, make it more difficult for the operator to recover costs. This reduces its investment capacity, resulting in a worsening economic scarcity. Water wastage has a direct impact on the balance between resource and demand, and the emergence of new adaptation behaviors on the consumer side introduces new factors to be taken into consideration in the overall supply management process.

This diagram represents the multidimensionality of the causal chain characteristic of intermittency, taking up the notions of global degradation expressed by the spiral and vicious circle of intermittency, while specifying the relationships between the different dimensions of the regime. In the following, we propose a methodology for quantifying this degradation through the definition of water service quality in the context of intermittency.

4. Framework Construction of Service Quality Indicators under IWS

4.1. Service Quality under IWS

Intermittent supply can be considered from the point of view of the consumer or that of water stakeholders. If we consider the consumer's point of view, the notion of intermittency can be categorized according to the level of standardized inconvenience generated, it is a categorization associated with the quality of the water delivery service. On the side of water stakeholders, it is rather an exploration of the reasons why this type of supply has been used, and their manifestation at the level of network management at all its stages.

The competitiveness of a business is primarily defined by the perceived quality of the product or service it offers [72]. In the following, we will explore the definition of water service quality as defined by the international institutions that outlined the Right to Water, and then compare it to the service perceived and received by consumers. The objective of this exercise is to assess the relevance of the traditional service quality criteria in IWS conditions, by comparing them with the reality of the situation experienced by the populations as described by the aforementioned problems.

4.2. The Right to Water and SDG Target

The United Nations Committee on Economic, Social and Cultural Rights (CESCR) defines the right to water as follows (2002):

"The human right to water entitles everyone to sufficient, safe, acceptable, physically accessible and affordable water for personal and domestic uses. An adequate amount of safe water is necessary to prevent death from dehydration, to reduce the risk of water-related disease and to provide for consumption, cooking, personal and domestic hygienic requirements" (CESCR, 2002).

The detailed explanation of the previous definition displays the five following parameters: availability; quantity; quality; affordability; and accessibility.

On the other hand, the Target 6.1 of the SDG is to achieve, by 2030, universal and equitable access to safe and affordable drinking water for all. The WHO/UNICEF Joint Monitoring Programme for Water Supply and Sanitation translated the terms used in this target into the same parameters for defining a ladder for monitoring household drinking water services [73].

Based on these criteria, service quality states can then be defined, and improvements can be noted in each of these axes. In the following paragraphs, we will proceed at representing service quality levels relating to each of these parameters, by adapting them to IWS conditions.

4.2.1. Availability

When the availability of water in an intermittent system is described, it is generally the average distribution time that is considered (in hours per day or per week . . .). Unfortunately, this indicator is not sufficient to describe the system, the frequency of distribution should also be considered [10], as well as its regularity [5]:

- Frequency or type of regime: Depending on the frequency of water arrival at homes or collection points, the service and accommodation needs of the populations may differ, so, with the same remaining parameters, a water supply of 3 h a day is different from a water supply every Monday, which in turn is different from a supply of once or twice a month.
- Regularity of service: this criterion categorizes the type of distribution as unreliable, predictable, regular [5], and continuous. *Unreliable* distribution characterizes a distribution regime in which consumers do not know exactly when they will receive water. When distribution is *regular*, the customer knows that he will get water a known number of times over a given period, without having an accurate knowledge of the exact day or time the water arrives. Finally, *predictable* service is one in which consumers know exactly which days and times the water arrives at their collection point. The last level of predictability is in continuous service, when the consumer no longer has to worry about when the water will be available.

Finally, another parameter that can be taken into account is the seasonality of water distribution, between a service that is permanently intermittent, and an intermittency that occurs during a particular season or as an isolated episode. This parameter can be integrated as a function of the evolution of the availability criterion with time, with peak factors calibrated to the impact of seasonality.

4.2.2. Quantity

The water needs of the consumer can be classified according to a pyramid [74] (see Figure 12) inspired by that of Maslow, starting from the basic water requirements related to human activities like those related to drinking, sanitation, bathing, and food preparation [75] and going upward to more flexible water uses.

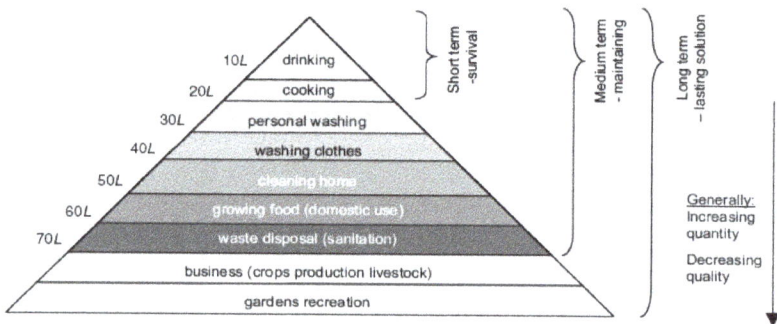

Figure 12. Hierarchy of water requirements (liter per capita per day) [74].

For our framework, we retrieve only the global aggregation of the average quantities associated with survival, maintenance, and development needs as an average proposition, considering that on the one hand, water is considered to be wasted under intermittent supply conditions on premises, while on the other hand some studies reveal that intermittent water supply of 1–6 h for example results in lower water use (30–60 L per capita per day —lpcd—) than continuous water supply (70 lpcd) [14].

4.2.3. Quality

Generally, water is either potable or non-potable, this binarity makes it difficult to qualify a progressive improvement in the quality of a given water. In this sense, the contractual provisions of networks under intermittent supply mean that the formulation of their water quality commitments may differ from conventional contracts. In the case of intermittency, the terms stipulate an obligation of drinkability, only at the exit of the water treatment plant, and in some other contracts, the consumer's water is qualified as "clean" rather than potable. While the criteria for defining this so-called "clean" water can be characterized on a case-by-case basis, there is no unanimous definition in this sense.

Indeed, the evolution of the field of water quality has made it possible to move from an exclusively organoleptic assessment to physico-chemical and microbiological criteria providing information on the possibility of its consumption [8].

The reality is that different uses of water do not necessarily require that it be potable to be used in activities other than drinking, cooking and, to some extent, hygiene, and this is where the concept of permissible risk comes in [76].

Based on this concept, the World Health Organization has defined a scale of health objectives related to microbiological risks, which represent the greatest threats to consumer safety (WHO, 2011). Thus, before being potable, water can be qualified according to its treatment objective, the World Health Organization has defined intermediate treatment objectives according to the admissible risk considered, namely: highly protective, protective, or interim [77] (Performance requirements for those criteria in Table 3). We adopt these levels in our scale, assuming a preliminary organoleptic qualification.

Table 3. Performance requirements for Household Water Treatment technologies and associated log10 reduction criteria for "interim", "protective", and "highly protective" risk protection targets [76].

Reference Microbe Used in Dose-Response Model	Assumed Number of Microbes Per Liter Used in Risk Calculations	Pathogen Class	Log10 Reduction Required		
			Interim	Protective	Highly Protective
			Requires Correct, Consistent and Continuous Use to Meet Performance Levels		
Campylobacter jejuni	1	Bacteria	Achieves "protective" target for two classes of pathogens and results in health gains	≥ 2	≥ 4
Rotavirus	1	Viruses		≥ 3	≥ 5
Cryptosporidium sp.	0.1	Protozoa		≥ 2	≥ 4

4.2.4. Accessibility

Water accessibility is defined by travel distance and waiting time, and the means needed to access water. Since the scope we considered to define intermittency only includes piped supply, it is the qualification of the delivery point that establishes the consumer's access to water.

Access can be improved from collective access, at kiosks or standpipes, to semi-collective access at a residence, neighborhood, or building, then to individual access.

This access also concerns alternative means of service (tanker trucks, water sales points, etc.).

However, discreet categorization cannot be perfectly objective since access to a collective terminal at the foot of one's home is better than access to a semi-collective delivery point that is further away.

The classification criterion is therefore a coefficient taking into account both the linear travelled and the average waiting time before recovery and consumption of this water (cf. Equation (1)).

$$\text{Water}_{\text{Accessibility}}(\%) = A \times \frac{\text{Linear}_{\text{travelled}}(m)}{1000} + B \times \frac{\text{Waiting}_{\text{time}}(min)}{30} \tag{1}$$

where 1000 m and 30 min are the maximum distance and time defined by the United Nations for Water fetching [78], and Where A and B are weighting coefficients that allow access rating (A can take into account the topography, and the available or required transport means, etc.). $A + B = 100$.

It should be noted that one of the issues of intermittency is when the consumer has a particular connection, but no water. This is not really included in accessibility, as it is defined here, the waiting time for water to reach the tap isn't considered. In our case, where people have tanks, there is no travel distance, and ultimately, no waiting time. The inconvenience that this situation causes to consumers is rather taken into account in the availability parameter.

4.2.5. Affordability

Water affordability is a vital element to water access, it relates to the ability of consumers to support the cost of water service. This parameter is typically measured by the annual cost of water bills as a percentage of median household income [79]. In Intermittent Water Supply conditions, the price of water has two components: the direct price of access to the service and the coping costs.

As far as water pricing is concerned, several models can be used depending on the socio-economic specificities of the water policies of the region, prices can be metering-based or lump-sum.

Coping costs can be higher than the actual price of piped water as is the case in Nepal households [46] (cf. Figure 13).

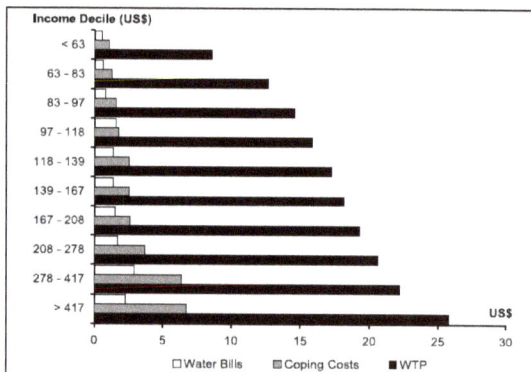

Figure 13. Household level mean monthly coping costs, water bill, and Willingness to Pay (WTP) by income decile in Nepal households [45].

An improvement in the quality of service must take these two components into consideration, the following formula (cf. Equation (2)) is a decomposition of Christian Smith's Water Affordability calculation [79]:

$$\text{Water}_{\text{Affordability}}(\%) = \frac{\text{Annual}_{\text{WaterServiceCost}} + \text{Annual}_{\text{WaterCopingCost}}}{\text{Annual}_{\text{HouseholdMedianIncome}}} \times 100 \qquad (2)$$

The graph above (Figure 14), shows the WTP values of the homes in Nepal, and displays the increase of this ratio with the decrease in income, another striking fact is that the willingness to pay for a better quality of service of all social categories is superior to the maximum threshold recommended by the United Nations (3%), hence the importance of establishing a scale of improvement of the affordability of water.

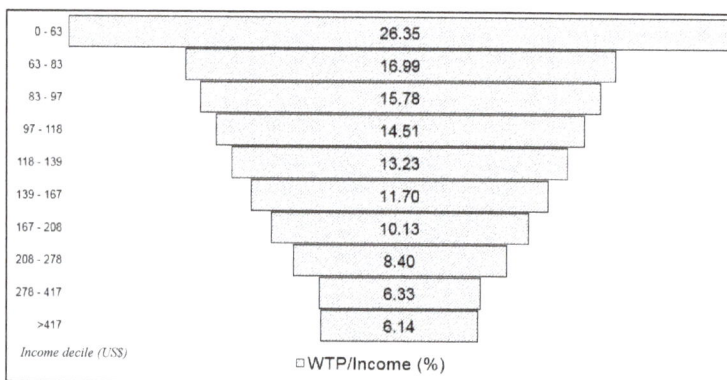

Figure 14. Willingness to Pay (WTP) per average revenue in Nepal households, adapted from Pattanayak's et al. graph (Figure 13).

4.3. Global Framework

In order to have an aggregated representation of water distribution service quality levels, the following diagram groups the five parameters described above by assigning ladders adapted to intermittent supply conditions.

The different levels represented in Figure 15, must be viewed independently from one column to another. They make it possible to identify opportunities for improvement and to qualify service levels. These service levels are associated in their final level with a continuous supply of good quality, fully satisfying the consumer's right to water, but the evidence shows that the existence of lower levels makes it possible to identify gradual improvement targets that can bring substantial benefit to those who will remain fed intermittently for a long time.

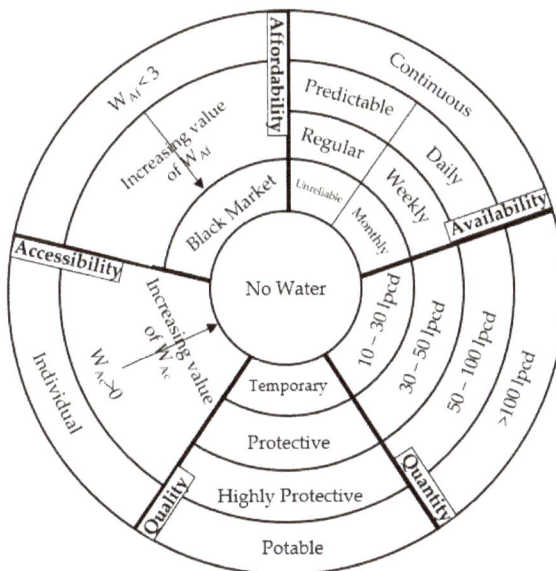

Figure 15. Classification of water service quality improvement framework.

We applied this general framework to the case of Saltillo's network (2600 km, 250,000 connections), and the considered parameters were as follows:

- Quantity: On a population base of 800,000, the allocation is of 113 L per person per day, people have enough water to cover their basic needs then (>100 lpcd).
- Quality: In 2014, 2740 samples of bacteriological analysis were performed together with authorities such as the Ministry of Health. All samples were in conformity with the standards [80]. But this analysis was performed at the level of the network and doesn't take into consideration the water contamination at the consumer's houses. In 2017, the water company received 74 water quality complaints (17 odor related, 39 due to the water's earthiness (53%), 14 because of the water color (19%), and four complaints associated with larvae presence (5%)). For a served population of 800,000 inhabitants, this figure stays very little even compared to continuous networks, if the water is not potable at the consumer's houses, the considered admissible risk places the quality of water as "highly protective" at the very least.
- Availability: The average water supply frequency is near 18.5 h/user/day, with 99% of the population receiving water every day and the rest 1% for 4 days a week. Moreover, according the water company, the distribution schedule is published by different means to make it possible to know when users, by neighborhood, are receiving water.

- Accessibility: Around 99% of the population is connected to the network, the accessibility is therefore individual.
- Affordability: the applied water tariffs deferred to annual consumption in Saltillo is on average 0.5% of the average annual income of the population, if we refer to the ratios calculated for the aforementioned coping costs in Nepal, the affordability indicator remains below the recommended threshold.

In short, the analysis of the quality of service in Saltillo places the majority of indicators at very favorable levels, but this is not the case throughout the network, which leads us to the importance of scaling up.

4.4. Scaling Up

The quality of service as defined above considers the response to a single person's water need, and yet it covers a wide spectrum of service improvement opportunities. However, other opportunities for improvement are felt as soon as we go beyond an individual scale.

Indeed, when faced with a supply that does not ensure water all the time. One of the vital aspects to take into account is to see if the water distribution is at least carried out in an equitable way. This includes both equity in the distribution of the quantities of water served, equity in terms of supply time, etc., and so on, until equity is considered in all the parameters identified above.

The criterion of equity, as well as other collective criteria can be grouped in the notion of regulation and regularization of the impact of a consumer or group of consumers on its environment, and the supplier's ability and willingness to trim, soften, or homogenize the spatial curve of supply quality.

The importance of this aspect is particularly visible in the case of the improvement of Jeddah's water distribution service. This improvement was realized thanks to the creation of adapted distribution sectors, along with water rationing upgrading using a new telecontrol center, and pressure modulation for water losses reduction. Let us take our framework to examine the various elements during the initial situation in May 2008, and after the improvement of the service:

- In terms of quantity, the water received by a Saudi household is more than sufficient, it even competes with American households' daily consumption of 315 lpcd [40].
- Water distribution was also predictable but with very disparate frequencies between areas of the city, water could in this sense arrive at some places every ten days, and at more than 27 days' interval in others, for an average frequency of 23 days.
- In terms of quality, the water company did not receive any complaints in this regard, we can deduce that the disadvantage generated by the degradation of water quality was not obvious in the organoleptic sense, but since it was not intended for drinking or cooking, this reduces the scope of impact it could have had on the population. It is, however, important to note that the long water storage time (23 days) necessarily degrades water quality.

Moreover, water accessibility is individual, and is subsidized by the state.

The most notable service improvement has been to act on this frequency, and on distribution equity, through active pressure management and the optimization of the network sectorization and supply schedules. The frequency of distribution increased from an initial average of once in 22.9 days to once in 7.2 days for the entire city in December 2014.

4.5. Service Quality Evaluation

The example of this network illustrates both the importance of considering equity in assessing service quality, and the subjectivity of improvement criteria according to the economic and social contexts of a network. Hence, the importance of adopting each network's qualification score according to its own improvement objectives, which are defined by the various parties concerned.

A qualification proposal can be formulated as a weighted sum of the impact of the various parameters detailed above (cf. Equation (3)):

$$SQ_{Score} = a.Avail_{Score} + b.Qt_{Score} + c.Ql_{Score} + d.Acc_{Score} + e.Aff_{Score} + f.Equ_{Score} \tag{3}$$

where: $a + b + c + d + e + f = 1$ and a, b, c, d, e, and f are weighting factors determined according to the networks improvement objectives.

5. Conclusions

The intermittency of drinking water distribution is characteristic of several countries in the world but there is no unanimous comprehensive definition of this regime. Despite the diversity of definitions associated with intermittency, there are constant elements that allow it to be established as a more or less regular time-bound water service, where the pressure at the exit points is not permanently above a given threshold.

It is difficult to give an exact history of intermittent initiation but the problems it generates are recurrent, even if their perception by operators and consumers is not the same, and the impact they have depends on the characteristics of the intermittency regime and on the socio-economic setting of the impacted partakers.

Intermittency is a relatively logical perpetuation of the current climatic, demographic and environmental conjecture, usually associated with organizational dysfunctions with multilevel contributions. There is a strong connection between the causes of intermittency and the problems associated with it. These relationships are globally reflected in a deterioration in the state of the system and in the quality of drinking water delivery service, leading to a worsening of the triggering factors, and an ingress into a characteristic vicious circle. However, the impairment can be scaled according to the inconvenience incurred.

The parameters defining service quality under intermittent water supply are adapted indicators of availability, quantity, quality, affordability, accessibility, and equity. The adaptation of those indicators to IWS conditions is done at the scaling level, by detailing the progress that can be measured.

An improvement in water service is possible even in cases where continuity is not achievable. The definition of improvement objectives and how they are perceived by stakeholders must be based on a targeted weighting of the parameters defining this quality of service and the potential contribution of the designated improvement in breaking the vicious circle of intermittency.

This work will be supplemented by methodological studies to define the objectives and means of improving each of the above-mentioned parameters, with the aim of defining an overall approach of resilient IWS, along with a characterization of the average weights for the qualification of these parameters through benchmarking field studies.

Author Contributions: A.M., B.T., A.C., B.d.G., and F.F. conceived and designed the methodology; A.C. and F.F. gathered the data; A.M. interviewed the operators and prepared the initial draft; all authors contributed to writing.

Funding: This research received no external funding except for the PhD grant partly funded by SUEZ Group.

Acknowledgments: This paper was written in the framework of a thesis partly funded by SUEZ Group.

Conflicts of Interest: The authors declare no conflicts of interest.

References

1. The United Nations. *The Human Right to Water and Sanitation-Media Brief*; The United Nations: Geneva, Switzerland, 2015; pp. 1–8.
2. World Health Organization. *Unicef Safely Managed Drinking Water-Thematic Report on Drinking Water*; World Health Organization: Geneva, Switzerland, 2017.

3. De Marchis, M.; Fontanazza, C.M.; Freni, G.; Loggia, G.L.; Napoli, E.; Notaro, V. Analysis of the impact of intermittent distribution by modelling the network-filling process. *J. Hydroinform.* **2011**, *13*, 358–373. [CrossRef]

4. Galaitsi, S.; Russell, R.; Bishara, A.; Durant, J.; Bogle, J.; Huber-Lee, A. Intermittent domestic water supply: A critical review and analysis of causal-consequential pathways. *Water* **2016**, *8*, 274. [CrossRef]

5. Charalambous, B.; Laspidou, C. *Dealing with the Complex Interrelation of Intermittent Supply and Water Losses*; IWA Publishing: London, UK, 2017.

6. Kumpel, E.; Nelson, K.L. Comparing microbial water quality in an intermittent and continuous piped water supply. *Water Res.* **2013**, *47*, 5176–5188. [CrossRef] [PubMed]

7. Philipp, K. Technical causes and impacts of intermittent water distribution. *Water Sci. Technol. Water Supply* **2012**, *12*, 504–512. [CrossRef]

8. World Health Organization. *Progress on Drinking Water, Sanitation and Hygiene*; World Health Organization: Geneva, Switzerland, 2017.

9. Kumpel, E. Water Quality and Quantity in Intermittent and Continuous Piped Water Supplies in Hubli-Dharwad, India. Ph.D. Thesis, University of California, Berkeley, CA, USA, 2013.

10. Guragai, B.; Takizawa, S.; Hashimoto, T.; Oguma, K. Effects of inequality of supply hours on consumers' coping strategies and perceptions of intermittent water supply in Kathmandu Valley, Nepal. *Sci. Total Environ.* **2017**, *599*, 431–441. [CrossRef] [PubMed]

11. Kumpel, E.; Nelson, K.L. Mechanisms affecting water quality in an intermittent piped water supply. *Environ. Sci. Technol.* **2014**, *48*, 2766–2775. [CrossRef] [PubMed]

12. Al-Ghamdi, A.S. Leakage—pressure relationship and leakage detection in intermittent water distribution systems. *J. Water Supply Res. Technol.* **2011**, *60*, 178–183. [CrossRef]

13. Rosenberg, D.E.; Talozi, S.; Lund, J.R. Intermittent water supplies: Challenges and opportunities for residential water users in Jordan. *Water Int.* **2008**, *33*, 488–504. [CrossRef]

14. Fan, L.; Liu, G.; Wang, F.; Ritsema, C.J.; Geissen, V. Domestic water consumption under intermittent and continuous modes of water supply. *Water Resour. Manag.* **2014**, *28*, 853–865. [CrossRef]

15. Whittington, D. Household demand for improved piped water services: Evidence from Kathmandu, Nepal. *Water Policy* **2002**, *4*, 531–556. [CrossRef]

16. Ayoub, G.M.; Malaeb, L. Impact of intermittent water supply on water quality in Lebanon. *Int. J. Environ. Pollut.* **2006**, *26*, 379–397. [CrossRef]

17. Agathokleous, A.; Christodoulou, S. The impact of intermittent water supply policies on urban water distribution networks. *Procedia Eng.* **2016**, *162*, 204–211. [CrossRef]

18. Charalambous, B. The hidden costs of resorting to intermittent supply. *Water* **2011**, *21*, 29–30.

19. Vairavamoorthy, K. Water Distribution Networks: Design and Control for Intermittent Supply. Ph.D. Thesis, Imperial College of Science and Technology, London, UK, 1994.

20. Andey, S.P.; Kelkar, P.S. Influence of intermittent and continuous modes of water supply on domestic water consumption. *Water Resour. Manag.* **2009**, *23*, 2555–2566. [CrossRef]

21. Taylor, R. *IWS: An International Update*; IWS: Waitakere city, Auckland, New Zealand, 2014.

22. Gohil, R.N. Continuous supply system against existing intermittent supply system. *Int. J. Sci. Res. Dev.* **2013**, *1*, 109–110.

23. Clark, R.M.; Hakim, S. *Securing Water and Wastewater Systems*; Springer Science & Business Media: Berlin, Germany, 2013; ISBN 978-3-319-01091-5.

24. Sashikumar, N.; Mohankumar, M.S.; Sridharan, K. Modelling an Intermittent Water Supply. In Proceedings of the World Water Environmental Resources Congress, Philadephia, PA, USA, 23–26 June 2003; pp. 1–11. [CrossRef]

25. Kumpel, E.; Nelson, K.L. Intermittent water supply: Prevalence, practice, and microbial water quality. *Environ. Sci. Technol.* **2016**, *50*, 542–553. [CrossRef] [PubMed]

26. Abu-Madi, M.; Trifunovic, N. Impacts of supply duration on the design and performance of intermittent water distribution systems in the West Bank. *Water Int.* **2013**, *38*, 263–282. [CrossRef]

27. McIntosh, A.C. *Asian Water Supplies: Reaching the Urban Poor: A Guide and Sourcebook on Urban Water Supplies in Asia for Governments, Utilities, Consultants, Development Agencies, and Nongovernment Organizations*; Asian Development Bank: Mandaluyong, Philippines; International Water Association: Manila, Philippines, 2003; ISBN 978-971-561-380-4.

28. Vacs Renwick, D.A. The Effects of an Intermittent Piped Water Network and Storage Practices on Household Water Quality in Tamale, Ghana. Ph.D. Thesis, Massachusetts Institute of Technology, Cambridge, MA, USA, 2013.

29. Solgi, M.; Bozorg Haddad, O.; Ghasemi Abiazani, P. Optimal intermittent operation of water distribution networks under water shortage. *J. Water Wastewater* **2016**, *28*. [CrossRef]

30. Ilaya-Ayza, A.E. Propuesta Para la Transición de un Sistema con Suministro de Agua Intermitente a Suministro Continuo. Ph.D. Thesis, Universitat Politècnica de València, València, Spain, 2016.

31. Brick, T.; Primrose, B.; Chandrasekhar, R.; Roy, S.; Muliyil, J.; Kang, G. Water contamination in urban south India: Household storage practices and their implications for water safety and enteric infections. *Int. J. Hyg. Environ. Health* **2004**, *207*, 473–480. [CrossRef] [PubMed]

32. Vairavamoorthy, K.; Gorantiwar, S.D.; Pathirana, A. Managing urban water supplies in developing countries—Climate change and water scarcity scenarios. *Phys. Chem. Earth Parts ABC* **2008**, *33*, 330–339. [CrossRef]

33. Bozorg-Haddad, O.; Hoseini-Ghafari, S.; Solgi, M.; Loáiciga, H.A. Intermittent urban water supply with protection of consumers' welfare. *J. Pipeline Syst. Eng. Pract.* **2016**. [CrossRef]

34. Sridhar, S. Intermittent Water Supplies: Where and Why They Are Currently Used and Why Their Future Use Should Be Curtailed. Ph.D. Thesis, University of Toronto, Toronto, ON, Canada, 2013.

35. Lieb, A.M.; Rycroft, C.H.; Wilkening, J. Optimizing intermittent water supply in urban pipe distribution networks. *SIAM J. Appl. Math.* **2016**, *76*, 1492–1514. [CrossRef]

36. Haddad, M.; Mcneil, L.; Omar, N. Model for Predicting Disinfection By-product (DBP) formation and occurrence in intermittent water supply systems: Palestine as a Case Study. *Arab. J. Sci. Eng.* **2014**, *39*, 5883–5893. [CrossRef]

37. CPHEEO. *Government of India Manual on Water Supply and Treatment*; GOI: New Delhi, India, 1999.

38. Simukonda, K.; Farmani, R.; Butler, D. Causes of water supply intermittency in Lusaka City in Zambia. *Water Pract. Technol.* **2018**, *13*, 335–345. [CrossRef]

39. Mohammed, A.B.; Sahabo, A.A. Water Supply and distribution problems in developing countries: A case study of Jimeta-Yola, Nigeria. *Int. J. Sci. Eng. Appl. Sci.* **2015**, *1*, 473–483.

40. Dieter, C.A.; Maupin, M.A. *Public Supply and Domestic Water Use in the United States, 2015*; U.S. Geological Survey: Reston, VA, USA, 2017.

41. Totsuka, N.; Trifunovic, N.; Vairavamoorthy, K. Intermittent urban water supply under water starving situations. In Proceedings of the 30th WEDC International Conference on Peoplecentered Approaches to Water and Environmental Sanitation, Vientiane, Lao PDR, 25–29 October 2004.

42. Soltanjalili, M.-J.; Bozorg Haddad, O.; Mariño, M.A. Operating water distribution networks during water shortage conditions using hedging and intermittent water supply concepts. *J. Water Resour. Plan. Manag.* **2013**, *139*, 644–659. [CrossRef]

43. Charalambous, B.; Nguyen, B. *The Challenge of Dealing with Intermittent Water Supply*; IWA: Tehran, Iran, 2017; p. 29.

44. Criminisi, A.; Fontanazza, M.; Freni, G.; Loggia, G.L. Evaluation of the apparent losses caused by water meter under-registration in intermittent water supply. *Water Sci. Technol. WST* **2009**, *60*, 2373–2382. [CrossRef] [PubMed]

45. de Marchis, M.; Fontanazza, C.M.; Freni, G.; Loggia, G.L.; Notaro, V.; Puleo, V. A mathematical model to evaluate apparent losses due to meter under-registration in intermittent water distribution networks. *Water Sci. Technol. Water Supply* **2013**, *13*, 914–923. [CrossRef]

46. Pattanayak, S.K.; Yang, J.-C. Coping with unreliable public water supplies-Averting expenditures by households in Kathmandu, Nepal. *Water Resour. Res.* **2005**, *41*. [CrossRef]

47. Cook, J.; Kimuyu, P.; Whittington, D. The costs of coping with poor water supply in rural Kenya: Coping costs of poor water. *Water Resour. Res.* **2016**, *52*, 841–859. [CrossRef]

48. De Marchis, M.; Fontanazza, C.M.; Freni, G.; La Loggia, G.; Napoli, E.; Notaro, V. A model of the filling process of an intermittent distribution network. *Urban Water J.* **2010**, *7*, 321–333. [CrossRef]

49. Marchis, M.D.; Fontanazza, C.M.; Freni, G.; Loggia, G.L.; Napoli, E.; Notaro, V. Modeling of distribution network filling process during intermittent supply. In Proceedings of the 10th Conference on Computing and Control for the Water Industry, Sheffield, UK, 19 September 2013.

50. Cabrera-Bejar, J.A.; Tzatchkov, V.G. Inexpensive modeling of intermittent service water distribution networks. *Am. Soc. Civ. Eng.* **2009**. [CrossRef]
51. Baisa, B.; Davis, L.W.; Salant, S.W.; Wilcox, W. The welfare costs of unreliable water service. *J. Dev. Econ.* **2010**, *92*, 1–12. [CrossRef]
52. Kumpel, E.; Woelfle-Erskine, C.; Ray, I.; Nelson, K.L. Measuring household consumption and waste in unmetered, intermittent piped water systems: Water use in intermittent piped systems. *Water Resour. Res.* **2017**, *53*, 302–315. [CrossRef]
53. Elala, D.; Labhasetwar, P.; Tyrrel, S.F. Tyrrel Deterioration in water quality from supply chain to household and appropriate storage in the context of intermittent water supplies. *Water Sci. Technol. Water Supply* **2011**, *11*, 400–408. [CrossRef]
54. Barrera, R.; Avilla, J.; Gonzalez-Tellez, S. Unreliable Supply of Potable Water and Elevated Aedes Aegypti Larval Indices: A Causal Relationship? *J. Am. Mosq. Control Assoc.* **1993**, *9*, 189–195. [PubMed]
55. Jensen, P.K.; Ensink, J.H.J.; Jayasinghe, G.; van der Hoek, W.; Cairncross, S.; Dalsgaard, A. Domestic transmission routes of pathogens: The problem of in-house contamination of drinking water during storage in developing countries. *Trop. Med. Int. Health* **2002**, *7*, 604–609. [CrossRef] [PubMed]
56. Bivins, A.W. Detailed Review of a Recent Publication: Intermittent water supply jeopardizes water quality and costs users and utilities money. *WaSH Policy Res. Dig.* **2017**, *7*, 1–4.
57. Bivins, A.W.; Sumner, T.; Kumpel, E.; Howard, G.; Cumming, O.; Ross, I.; Nelson, K.; Brown, J. Estimating infection risks and the global burden of diarrheal disease attributable to intermittent water supply using QMRA. *Environ. Sci. Technol.* **2017**, *51*, 7542–7551. [CrossRef] [PubMed]
58. Adane, M.; Mengistie, B.; Medhin, G.; Kloos, H.; Mulat, W. Piped water supply interruptions and acute diarrhea among under-five children in Addis Ababa slums, Ethiopia: A matched case-control study. *PLoS ONE* **2017**, *12*, e0181516. [CrossRef] [PubMed]
59. Dunkle, S.E.; Mba-Jonas, A.; Loharikar, A.; Fouché, B.; Peck, M.; Ayers, T.; Archer, W.R.; de Rochars, V.M.B.; Bender, T.; Moffett, D.B.; et al. Epidemic cholera in a crowded urban environment, Port-au-Prince, Haiti. *Emerg. Infect. Dis. J.* **2011**, *17*. Available online: http://sci-hub.tw/https://wwwnc.cdc.gov/eid/article/17/11/11-0772_article (accessed on 28 June 2018). [CrossRef] [PubMed]
60. Nyende-Byakika, S.; Ngirane-Katashaya, G.; Ndambuki, J.M. Modeling flow regime transition in intermittent water supply networks using the interface tracking method. *Int. J. Phys. Sci.* **2012**, *7*, 327–337. [CrossRef]
61. Batish, R. A new approach to the design of intermittent water supply networks. In Proceedings of the World Water and Environmental Resources Congress, Philadelphia, PA, USA, 23–26 June 2003.
62. Christodoulou, S.; Agathokleous, A. A study on the effects of intermittent water supply on the vulnerability of urban water distribution networks. *Water Sci. Technol. Water Supply* **2012**, *12*, 523–530. [CrossRef]
63. Mastaller, M.; Klingel, P. Application of a water balance adapted to intermittent water supply and flat-rate tariffs without customer metering in Tiruvannamalai, India. *Water Sci. Technol. Water Supply* **2018**, *18*, 347–356. [CrossRef]
64. Ilaya-Ayza, A.; Martins, C.; Campbell, E.; Izquierdo, J. Implementation of DMAs in intermittent water supply networks based on equity criteria. *Water* **2017**, *9*, 851. [CrossRef]
65. Freni, G.; De Marchis, M.; Napoli, E. Implementation of pressure reduction valves in a dynamic water distribution numerical model to control the inequality in water supply. *J. Hydroinform.* **2014**, *16*, 207. [CrossRef]
66. Ameyaw, E.E.; Memon, F.A.; Bicik, J. Improving equity in intermittent water supply systems. *J. Water Supply Res. Technol. AQUA* **2013**, *62*, 552–562. [CrossRef]
67. Solgi, M.; Bozorg Haddad, O.; Seifollahi-Aghmiuni, S.; Loáiciga, H.A. Intermittent operation of water distribution networks considering equanimity and justice principles. *J. Pipeline Syst. Eng. Pract.* **2015**. [CrossRef]
68. Vairavamoorthy, K.; Ali, M. optimal design of water distribution systems using genetic algorithms. *Comput. Aided Civ. Infrastruct. Eng.* **2000**, *15*, 374–382. [CrossRef]
69. Gottipati, P.V.K.S.V.; Nanduri, U.V. Equity in water supply in intermittent water distribution networks: Equity in water supply. *Water Environ. J.* **2014**, *28*, 509–515. [CrossRef]
70. Taylor, D.D.J. Reducing Booster-Pump-Induced Contaminant Intrusion in Indian Water Systems with a Self-Actuated, Back-Pressure Regulating Valve. Ph.D. Thesis, Massachusetts Institute of Technology, Cambridge, MA, USA, 2014.

71. Chary, S. *Real Cost of Intermittent Water Supply for Consumers in India*; Global Water Intelligence: Oxford, UK, 2009; Volume 10.

72. Mosahab, R. Service quality, customer satisfaction and loyalty: A test of mediation. *Int. Bus. Res.* **2010**, *3*, 72. [CrossRef]

73. WHO. *Potable Resuse: Guidance for Producing Safe-Drinking Water*; WHO: Geneva, Switzerland, 2017.

74. WHO Regional Office for South-East Asia. *Minimum Water Quantity Needed for Domestic Use*; Mahatma Gandhi Marg: New Delhi, India, 2005.

75. Gleick, P.H.; Iwra, M. Basic water requirements for human activities: Meeting basic needs. *Water Int.* **1996**, *21*, 83–92. [CrossRef]

76. Hunter, P.R.; Fewtrell, L. Acceptable Risk. In *Water Quality: Guidelines, Standards, and Health: Assessment of Risk and Risk Management for Water-Related Infectious Disease*; Fewtrell, L., Bartram, J., Eds.; World Health Organization: Geneva, Switzerland, 2001; ISBN 978-1-900222-28-0.

77. OMS. *Evaluation des Options de Traitement Domestique de L'eau*; Organisation Mondiale de la Santé, World Health Organization: Lyon, France, 2012.

78. United Nations Office to support the International Decade for Action "Water for life" 2005–2015. *UN-Water Decade Programme on Advocacy and Communication (UNW-DPAC)*; Casa Solans: Zaragoza, Spain, 2012.

79. Christian-Smith, J.; Balazs, C.; Heberger, M.; Longley, K.E. *Assessing Water Affordability: A Pilot Study in Two Regions of California*; Pacific Institute: Oakland, CA, USA, 2013; ISBN 978-1-893790-04-9.

80. Aguas de Saltillo City of Saltillo, Coahuila México. Available online: http://www.aquafed.org/Public/Files/__Uploads/files/Saltillo.pdf (accessed on 28 June 2018).

MDPI

St. Alban-Anlage 66

4052 Basel

Switzerland

Tel. +41 61 683 77 34

Fax +41 61 302 89 18

www.mdpi.com

Water Editorial Office

E-mail: water@mdpi.com

www.mdpi.com/journal/water

www.ingramcontent.com/pod-product-compliance
Lightning Source LLC
Chambersburg PA
CBHW051859210326
41597CB00033B/5961